Lecture Notes in Mathematics

Edited by A. Dold and B. Eckmann

Series: Department of Mathematics, University of Maryland,
College Park
Adviser: L. Greenberg

458

Peter Walters

Ergodic Theory –
Introductory Lectures

Springer-Verlag Berlin Heidelberg GmbH 1975

Dr. Peter Walters
Mathematics Institute
University of Warwick
Coventry/England

Library of Congress Cataloging in Publication Data

Walters, Peter, 1943–
    Ergodic theory.

    (Lecture notes in mathematics ; 458)
    Bibliography: p.
    Includes index.
    1. Ergodic theory. I. Title. II. Series:
Lectures notes in mathematics (Berlin) ; 458.
QA3.L28 no. 458 [QA313]   510'.8s [515'.42] 75-9853
ISBN 978-3-540-07163-1

AMS Subject Classifications (1970): 28 A 65

ISBN 978-3-540-07163-1   ISBN 978-3-540-37494-7 (eBook)
DOI 10.1007/978-3-540-37494-7

© by Springer-Verlag Berlin Heidelberg 1975.
Originally published by Springer-Verlag Berlin Heidelberg New York in 1975
Offsetdruck: Julius Beltz, Hemsbach/Bergstr.

Preface

These are notes of a one-semester introductory course on Ergodic
Theory that I gave at the University of Maryland in College Park
during the fall of 1970. I assumed the audience had no previous
knowledge of Ergodic Theory. My aim was to present some of the basic
facts in measure theoretic Ergodic Theory and Topological Dynamics and
show how they are related so that the audience would have the foundations to read the research papers if they wished to pursue the subject
further.

At the beginning of Chapter 1 I give a list of examples of
measure-preserving transformations and at the end of each section of
Chapter 1 I investigate whether these examples have the properties
discussed in that section. These examples were chosen because of
their varied properties and importance in the subject. Similarly in
Chapter 5, on Topological Dynamics, a list of examples is given and
the properties discussed in that chapter are considered for these
examples.

I tried to deal with entropy as simply as possible. In the discussion of entropy I have inserted without proof some of the more
difficult theorems when I thought they were relevant to the discussion.
In particular I have discussed the recent deep results of D. S.
Ornstein on Bernoulli automorphisms and Kolmogorov automorphisms.

In the final chapter I have presented the new treatment of topological entropy due to R. E. Bowen. One of the beauties of this
treatment is that topological entropy can be defined for a uniformly
continuous map of any metric space and that its value remains un-
changed under certain types of covering maps. This enables one to
give an elegant calculation of the topological and (Haar) measure

theoretic entropies of affine transformations of finite-dimensional tori.

Since these notes have not been fully edited many references are missing and it is likely that credit is often not given where it is due. The theorems and definitions are numbered independently, but a corollary is given the same number as the theorem to which it is a corollary.

Thanks are due to Victor Charles Stasio and Suellen Eslinger who took notes of the course and also to Allan Jaworski for editing and compiling the bibliography. Special thanks are due to Betty Vander-slice for her superb typing.

--Peter Walters

Contents

Chapter 0: Preliminaries

§1. Introduction

Generally speaking, ergodic theory is the study of transformations and flows from the point of view of recurrence properties, mixing properties, and other global, dynamical properties connected with asymptotic behavior. Abstractly, one has a space X and a transformation T of X (or a family $\{T_t : t \in \mathbb{R}\}$ of transformations of X) with some structure on X which is preserved by T (or by $\{T_t\}$). The nature of most of the work so far can be categorized into one of the four following types:

(1) measure theoretic:

Here one deals with a measure space X and a measure preserving transformation $T: X \rightarrow X$.

(2) topological:

Here X is a topological space and $T: X \rightarrow X$ is a continuous map.

(3) mixture of (1) and (2):

In this situation X is a topological space equipped with a measure m on its Borel sets while $T: X \rightarrow X$ is continuous and preserves m.

(4) smooth:

One considers a smooth manifold X and a smooth map $T: X \rightarrow X$.

We shall deal with some topics from (1), (2), and (3).

To see how this study arose consider, for example, a system of k particles in 3-space moving under known forces. Suppose that the phase of the system at a given time is completely determined by the

positions and the momenta of each of the k particles. Thus, at a
given time the system is determined by a point in 6k-dimensional
space. As time continues the phase of the system alters according to
the differential equations governing the motion, e.g., Hamilton's
equations

$$\frac{dq_i}{dt} = \frac{\partial H}{\partial p_i} \quad , \quad \frac{dp_i}{dt} = - \frac{\partial H}{\partial q_i} \quad .$$

If we are given an initial condition and such equations can be uniquely solved then the corresponding solution gives us the entire history
of the system, which is determined by a curve in phase space.

   If x is a point in phase space representing the system at a
time t_0, let $T_t(x)$ denote the point of phase space representing
the system at time $t+t_0$. From this we see that T_t is a transformation of phase space and, moreover, T_0 = id., $T_{t+s} = T_t \circ T_s$. Thus
$\{T_t: t \in \mathbb{R}\}$ is a one-parameter group of transformations of phase
space. In dynamics one is interested in the asymptotic properties of
the family $\{T_t\}$. It seems reasonable to study the system at discrete
times t_0, $2t_0$, $3t_0$,..., i.e., study the family $\{T_{t_0}^n\}_{n=1}^{\infty}$, since we
expect the properties of $\{T_t\}$ to be reflected in those of $\{T_{t_0}^n\}$.

   For this reason, as well as the fact that it is simpler, one
studies individual transformations and their iterates. One is par-
ticularly interested in the flow on an energy surface, which is sometimes smooth (hence considerations of type (4) arise), and sometimes
is not (one then investigates along the lines of (2)). Measure
theory enters the picture via a theorem of Liouville which tells us
that if the forces are of a certain type one can choose coordinates
in phase space so that the usual 6k-dimensional measure in these
coordinates is preserved by each transformation T_t.

   Around 1900 Gibbs suggested using the measure-theoretic approach

in mechanics because of the difficulty of solving the equations of
motion and also because this deterministic approach does not answer
several important questions in mechanics. In discovering statistical
mechanics Gibbs suggested looking at what happens to subsets of phase
space. For example, if A and B are subsets of phase space what
is the probability that the system is in B at the time t given
that the system is in A at the time t_0? Given that the system be-
gins in A at time t_0 what is the average time the system spends
in B? Such questions motivate the type of study undertaken in
ergodic theory.

We now list some general references for the material we shall
discuss:

For topics of types (1) and (3) mentioned above see Halmos [2],
Billingsley [1], Hopf [1], Jacobs [1] [2], Parry [3], Rohlin [3][4][5],
Friedman [1], Shields [2]. In addition to the Shields notes, further
details on the results of Ornstein described in Chapter 4 may be found
in a forthcoming book by Friedman and Ornstein [2].

For material of type (2) see Gottschalk and Hedlund [1],
Nemytskii and Stepanov [1], and Ellis [1].

And material of type (4) may be found in Avez and Arnold [1],
Smale [1], Abraham [1], Abraham and Robbin [1], Nitecki [1].

Khinchin [1] provides a good sketch of the foundations of ergodic
theory. For extensive bibliographies see Jacobs [1][2], Gottschalk
[1], and Smale [1]. A recent survey is Mackey [1].

§2. Measure Theory
General reference - Halmos [1].

We recall some fundamental notions from measure theory.

Let X be a set. A σ-algebra of subsets of X is a collection
B of subsets of X satisfying:

(1) $X \in \mathcal{B}$,   (2) $B \in \mathcal{B} \Rightarrow X \backslash B \in \mathcal{B}$,

(3) $B_n \in \mathcal{B}$, $n > 1 \Rightarrow \bigcup_{n=1}^{\infty} B_n \in \mathcal{B}$.

We then call $(X, \mathcal{B})$ a _measurable_ _space_. A _measure_ _space_ is a triple $(X, \mathcal{B}, m)$ where X is a set, $\mathcal{B}$ is a σ-algebra of subsets of X, and m is a function $m \colon \mathcal{B} \to R^{+}$ satisfying

$$ m\left(\bigcup_{n=1}^{\infty} B_n \right) = \sum_{n=1}^{\infty} m(B_n) $$

if $\{B_n\}$ is a pairwise disjoint sequence of elements of $\mathcal{B}$. We say that $(X, \mathcal{B}, m)$ is a _probability_ _space_, or a _normalized_ _measure_ _space_, if $m(X) = 1$. We shall usually deal with such spaces.

A collection A of subsets of a set X is an _algebra_ if:

(1) $X \in A$,   (2) $A \in A \Rightarrow X \backslash A \in A$,

(3) $A_1, \ldots, A_n \in A \Rightarrow \bigcup_{i=1}^{n} A_i \in A$.

When one is trying to equip a measurable space $(X, \mathcal{B})$ with a measure one usually knows what the measure should be on an algebra $A \subseteq \mathcal{B}$, and so, one would like to know when this function defined on A can be extended to a measure on $\mathcal{B}$. The following result deals with this situation.

Hahn-Kolmogorov _Extension_ _Theorem_:

Given a set X, an algebra A of subsets of X, let $m \colon A \to R^{+}$ be a function satisfying

$$ m(X) = 1, \quad m\left(\bigcup_n A_n \right) = \sum_n m(A_n) $$

whenever $A_n \in A \ \forall n$, $\bigcup_n A_n \in A$, and the $\{A_n\}$ are disjoint. Then there is a unique probability measure $\bar{m}$ defined on the σ-algebra generated by A such that $\bar{m}(A) = m(A)$ whenever $A \in A$.

A <u>monotone</u> <u>class</u> of subsets of X is a collection C of subsets of X such that if $E_1 \subset E_2 \subset \ldots$ belong to C then $\bigcup_n E_n \in C$ and if $F_1 \supset F_2 \supset \ldots$ belong to C then $\bigcap_n F_n \in C$.

<u>Theorem</u>:

If A is an algebra of subsets of X then the σ-algebra generated by A equals the monotone class generated by A.

If (X, B, m) is a finite measure space, one can easily deduce from this theorem that if A is an algebra generating the σ-algebra B then for any $B \in B$ and $\varepsilon > 0$ there exists $A \in A$ with $m(A \Delta B) < \varepsilon$. (In fact,

$$C = \{B \in B \mid \forall \ \varepsilon > 0 \ \ \exists \ B_\varepsilon \in A \ \ \text{with} \ \ m(B \Delta B_\varepsilon) < \varepsilon\}$$

is a monotone class and contains A.)

<u>Notation</u>: If A is an algebra we shall write $\sigma(A)$ for the σ-algebra generated by A.

<u>Direct</u> <u>Products</u>:

Let (X_i, B_i, m_i), $i \in Z$ be probability spaces. Their <u>direct</u> <u>product</u>

$$(X, B, m) = \prod_{i=-\infty}^{\infty} (X_1, B_1, m_1)$$

is defined as follows:

(a) $X = \prod_{i=-\infty}^{\infty} X_i$

(b) Let $n_1 < n_2 < \ldots < n_r$ be integers, and $A_{n_i} \in B_{n_i}$ $i = 1, \ldots, r$. We define a measurable rectangle to be a set of the form

$$\{(x_j) \in X: x_{n_i} \in A_{n_i} \ \forall \ i: \ 1 \le i \le r\}.$$

Let A be the collection of all finite unions of measurable rectangles. A is an algebra: (1) and (3) are obvious; to show (2) observe that

$$X \setminus \{(x_j) \mid x_{n_i} \in A_{n_i}, \quad 1 \le i \le r\} = \bigcup_{i=1}^{r} \{(x_j) \mid x_{n_i} \in X_{n_i} \setminus A_{n_i}\} \in A$$

and that A is closed under finite intersections. Let B be the σ-algebra generated by A.

(c) Each element of A can be written as a disjoint finite union of measurable rectangles so that we define

$$m(\{(x_j) \mid x_{n_i} \in A_{n_i}, \quad 1 \le i \le r\}) = \prod_{i=1}^{r} m_{n_i}(A_{n_i})$$

and then extend m to A in the obvious manner. The conditions of the Hahn-Kolmogorov Theorem can be shown to be satisfied, and thus we can extend m to B. Hence, we obtain a probability space.

Measurable Functions:

A Borel subset of R is a member of the σ-algebra generated by the open sets.

Let (X, B, μ) be a measure space; $f: X \to R$ is measurable if for all $c \in R$, $f^{-1}(c, \infty) \in B$, or equivalently $f^{-1}(D) \in B$ $\forall$ Borel sets $D \subset R$.

A function $f: X \to C$, $f = f_1 + if_2$ is measurable if $f_1: X \to R$ and $f_2: X \to R$ are measurable.

If X is a topological space and B is the σ-algebra of Borel subsets of X (the σ-algebra generated by the open subsets of X) then each continuous function $f: X \to C$ is measurable.

Integration:

A simple function is a function of the form $\sum_{i=1}^{n} a_i X_{A_i}$, where $a_i \in R$, the $\{A_i\}$ are disjoint members of B, and X_{A_i} denotes the characteristic function of A_i. Simple functions are measurable. We define the integral for simple functions by:

$$\int (\sum_{i=1}^{n} a_i X_{A_i}) dm = \sum_{i=1}^{n} a_i m(A_i).$$

Suppose f: X → R is measurable and f ≥ 0; then there exists
an increasing sequence of simple functions $f_n \nearrow f$. For example we
could take

$$
f_n(x) = \begin{cases} \dfrac{i-1}{2^n} & \text{if } \dfrac{i-1}{2^n} \le f(x) < \dfrac{i}{2^n} \quad i = 1,\ldots,n2^n \\[3mm] n & \text{if } f(x) \ge n. \end{cases}
$$

We define $\int f\, dm = \lim\limits_{n\to\infty} \int f_n\, dm$ and note that this definition is independent of the chosen sequence $\{f_n\}$.

Suppose f: X → R is measurable; then $f = f^+ - f^-$ where

$$
f^+(x) = \max\{f(x),0\} \ge 0
$$

$$
f^-(x) = \max\{-f(x),0\} \ge 0.
$$

We say that f is <u>integrable</u> if $\int f^+ dm,\ \int f^- dm < \infty$, and we define

$$
\int f\, dm = \int f^+ dm - \int f^- dm.
$$

Now if f: X → C is measurable, $f = f_1 + if_2$, f is <u>integrable</u> if
f_1 and f_2 are integrable and we define

$$
\int f\, dm = \int f_1\, dm + i \int f_2\, dm.
$$

Observe that f is integrable if and only if |f| is integrable.
Let $L^1(\mu)$ denote the class of all integrable functions (X,B,μ) → C.
<u>Lebesgue</u> <u>Dominated</u> <u>Convergence</u> <u>Theorem</u>:

Suppose $\{f_n\}$ is a sequence of measurable functions, $f_n \to f$ a.e.
and there exists an integrable function g such that $|f_n(x)| \le g(x)$
a.e. for all n; then f and each f_n are integrable and

$$
\int f_n\, d\mu \ \to\ \int f\, d\mu.
$$

§3. Hilbert Spaces

General reference - Halmos [3].

H is a Hilbert space if it is a Banach inner product space,
i.e., (1) H is a vector space over the complex numbers C

(2) There is an inner product on H, i.e., a map
$(\cdot,\cdot)$: $H \times H \to C$ such that:

(a) $(\cdot,\cdot)$ is bilinear

(b) $(f,f) \geq 0$ for all f in H

(c) $(f,f) = 0$ if and only if $f = 0$

(d) $(f,g) = \overline{(g,f)}$.

(3) $\|f\| = (f,f)^{\frac{1}{2}}$ is a norm on H inducing a complete metric
topology on H.

Let (X,B,m) be a measure space. Consider all measurable func-
tions $f: X \to C$ such that $\int |f|^2 dm < \infty$; we define an equivalence
relation on this set by saying that $f \sim g$ if and only if $f = g$ a.e.
The set of equivalence classes relative to this relation forms a Hilbert space which we denote by $L^2(X,B,m) = L^2(m)$, where the inner
product of two functions is given by

$$(f,g) = \int f \bar{g} \, dm.$$

Recall: The Schwarz Inequality

In any Hilbert space H

$$|(f,g)| \leq \|f\| \cdot \|g\| \quad \text{for all } f,g \text{ in } H.$$

A unitary operator U on a Hilbert space H is an isomorphism
of H, i.e., U is a linear bijective map preserving the inner
product $((Ux,Uy) = (x,y), x,y$ in $H)$. It follows that U is continuous.

§4. Haar Measure

General reference - Pontrjagin [1].

Theorem:

Let G be a compact topological group; then there exists a finite measure m defined on the Borel subsets of G such that $m(xE) = m(E)$ for all x in G, and for all Borel sets E. We call such a measure a Haar measure.

E.g., let $K = \{z \in C: |z| = 1\}$ and m denote normalized circular Lebesgue measure. Then $m(aU) = m(U)$ for all sets U measurable on K.

Theorem:

If m and μ are both finite Haar measures on the compact topological group G then $m = c\mu$ where $c > 0$. Thus there exists only one normalized Haar measure on G.

Remarks:

(1) If $U \subset G$ is a non-empty open set then it has positive Haar measure. This is because

$$G = \bigcup_{g \in G} gU = g_1 U \cup g_2 U \cup \ldots \cup g_n U$$

by compactness.

(2) In the Hilbert space $L^2(K, \mathcal{B}, m)$ where m is the Haar measure on the unit circle K, the functions

$$f(z) = z^n, \quad n \in Z$$

form an orthonormal basis.

§5. Character Theory

General reference - Pontrjagin [1].

Many of our examples will be rotations, endomorphisms or affine transformations of compact groups. (We mean endomorphism in the sense of topological groups, i.e., an abstract group endomorphism which is continuous.) In some proofs we will use the character theory of compact abelian groups, which we summarize in this section. For those not familiar with character theory, proofs in the later sections involving characters will usually be preceded by the proof in a special case where the group used is the unit circle and then classical Fourier analysis will be used.

Let G be a locally compact abelian group with a countable topological base. Let $\hat{G}$ denote the collection of all continuous homomorphisms of G into the unit circle K. The members of $\hat{G}$ are the characters of G. $\hat{G}$ is an abelian group under the operation of pointwise multiplication of functions. With the compact open topology $\hat{G}$ becomes a locally compact abelian group with a countable topological base. We have the following results:

(1) G is compact $\Leftrightarrow$ $\hat{G}$ is discrete.

(2) $(\hat{\hat{G}})$ is naturally isomorphic (as a topological group) to G, the isomorphism being given by the map

$$\alpha \to a \quad \text{where} \quad \alpha(\gamma) = \gamma(a) \quad \text{for all} \quad \gamma \in \hat{G}.$$

(3) If G is compact then G is connected $\Leftrightarrow$ $\hat{G}$ is torsion free.

(4) $\widehat{G_1 \times G_2} = \hat{G}_1 \times \hat{G}_2$ where "×" denotes direct product.

So in some sense we can study compact abelian groups by studying discrete countable groups.

Examples: (See §6 for proofs.)

(i) Let $G = K = \{z \in C: |z| = 1\}$. Each homomorphism of K to itself is of the form $z \to z^n$ $(n \in Z)$ so that $\hat{G} \simeq Z$.

(ii) Let $G = K^n$ the n-torus. By (4) $\hat{K}^n \simeq Z^n$ and in fact each member of $\hat{K}^n$ has the form

$$\gamma(z_1, z_2, \ldots, z_n) = z_1^{p_1} z_2^{p_2} \ldots z_n^{p_n}$$

where $(p_1, \ldots, p_n) \in Z^n$.

(5) If H is a closed subgroup of G and $H \neq G$ there exists a $\gamma \in \hat{G}$, $\gamma \neq 1$ such that $\gamma(h) = 1$ $\forall h \in H$. (We shall write this $\gamma(H) = 1$.)

(6) Let G be compact. The members of $\hat{G}$ are mutually orthogonal members of $L^2(m)$, where m is Haar measure.

Proof. It suffices to show

$$\int_G \gamma(x) dm(x) = 0 \quad \text{if} \quad \gamma \neq 1.$$

If $a \in G$ then since m is Haar measure

$$\int \gamma(x) dm(x) = \int \gamma(ax) dm(x) = \gamma(a) \int \gamma(x) dm(x).$$

Choosing a so that $\gamma(a) \neq 1$ we have $\int \gamma(x) dm(x) = 0$. //

(7) If G is compact, the members of $\hat{G}$ form an orthonormal basis for $L^2(m)$ where m is normalized Haar measure.

This is part of the Peter-Weyl theorem and can be easily deduced from the Stone-Weierstrass theorem, which implies that finite linear combinations of characters are dense in $C(G)$ = space of complex-valued continuous functions of G.

(8) If $A: G \to G$ is an endomorphism we can define an endomorphism $\hat{A}: \hat{G} \to \hat{G}$ by $\hat{A}\gamma = \gamma \circ A$, $\gamma \in \hat{G}$. It is easy to see that A is one-to-one if and only if $\hat{A}$ is onto and A is onto if and only if $\hat{A}$ is one-to-one. Therefore A is an automorphism if and only if $\hat{A}$ is an automorphism.

Recall that for compact groups G, G is metric iff G has a countable topological base.

§6. Endomorphisms of Tori

We shall view the n-torus in two ways:- multiplicatively as $K^n = \underbrace{K \times K \times \ldots \times K}_{n \text{ times}}$, and additively as R^n/Z^n where R^n is n-space and Z^n is the subgroup of R^n consisting of points with integer coordinates. A topological group isomorphism is given by $K^n \to R^n/Z^n$,

$$(e^{2\pi i x_1}, \ldots, e^{2\pi i x_n}) \mapsto (x_1, \ldots, x_n) + Z^n.$$

Theorem:

(1) Every closed subgroup of K is either K or is a finite cyclic group consisting of all p-th roots of unity for some integer $p > 0$.

(2) The only automorphisms of K are the identity and the map $z \mapsto z^{-1}$.

(3) The only homomorphisms of K are the maps

$$\phi_n: z \mapsto z^n, \quad n \in Z.$$

(4) The only homomorphisms of K^n to K are maps of the form

$$(z_1, \ldots, z_n) \mapsto z_1^{m_1} \cdot \ldots \cdot z_n^{m_n} \text{ where } m_1, \ldots, m_n \in Z.$$

Proof: (1) Let H be a closed subgroup of K; if H is infinite it has a limit point so, $\forall \varepsilon > 0 \; \exists \, a,b \in H \; \ni d(a,b) < \varepsilon$, $a \neq b$. Then $d(b^{-1}a, 1) < \varepsilon$ and therefore the elements of H are ε-dense in $K \; \forall \varepsilon$. Thus, $H = K$.

If H is finite and has p elements then $a^p = 1 \; \forall \, a \in H$. So each element of H is a p-th root of unity, and since there are p elements in H, H must consist of all the p-th roots of unity.

(2) Let $\theta: K \to K$ be an automorphism. $\theta(1) = 1$. Since -1 is the only element of K of order 2 we have $\theta(-1) = -1$. Since $i, -i$

are the only elements of order 4 either $\theta(i) = i$ and $\theta(-i) = -i$ or $\theta(i) = -i$ and $\theta(-i) = i$. Consider the first case. Since θ maps intervals to intervals, the interval $\overrightarrow{[1,i]}$ from 1 to i is either mapped to itself or to $\overrightarrow{[i,1]}$ (all intervals go anticlockwise). But since $\overrightarrow{[1,i]}$ does not contain -1 it cannot be mapped to $\overrightarrow{[i,1]}$ so $\theta\overrightarrow{[1,i]} = \overrightarrow{[1,i]}$. The only element of order 8 in $\overrightarrow{[1,i]}$ is $e^{\pi i/4}$ and so this must be fixed by θ. Therefore $\theta\overrightarrow{[1,e^{\pi i/4}]} = \overrightarrow{[1,e^{\pi i/4}]}$. By induction one shows that $\theta(e^{2\pi i/2^k}) = e^{2\pi i/2^k}$ for each $k > 0$. It follows that θ fixes all the 2^k-th roots of unity $\forall\, k > 0$ and hence is the identity. In the second case one shows that $\theta(e^{2\pi i/2^k}) = e^{-2\pi i/2^k}$ $\forall\, k > 0$ and hence $\theta(z) = z^{-1}$, $z \in K$.

(3) Let $\theta: K \to K$ be an endomorphism. If θ is non-trivial, its image, $\theta(K)$, is a closed connected subgroup of K and so $\theta(K) = K$ by (1). The kernel $\text{Ker }\theta$ is a closed subgroup of K so either $\text{Ker }\theta = K$ or $\text{Ker }\theta = H_p$, the group of all p-th roots of unity, for some p. The first case corresponds to trivial θ. If $\text{Ker }\theta = H_p$ let $\alpha_p: K/H_p \to K$ be the isomorphism given by $\alpha_p(zH_p) = z^p$, and let $\theta_1: K/H_p \to K$ be the isomorphism induced by θ ($\theta_1(zH_p) = \theta(z)$). Then $\theta_1\alpha_p^{-1}$ is an automorphism of K and by (2) either $\theta_1\alpha_p^{-1}(z) = z$ $\forall\, z \in K$ or $\theta_1\alpha_p^{-1}(z) = z^{-1}$ $\forall\, z \in K$. Hence either $\theta(z) = \theta_1(zH_p) = \theta_1\alpha_p^{-1}(z^p) = z^p$ $\forall\, z \in K$ or $\theta(z) = z^{-p}$ $\forall\, z \in K$.

(4) Let $\gamma_i: K \to K^n$ be defined by $\gamma_i(z) = (1,1,\ldots,1,z,1,\ldots,1)$.
$$\underset{\text{i-th place}}{\uparrow}$$
If $\theta: K^n \to K$ is a homomorphism then $\theta\circ\gamma_i: K \to K$ is an endomorphism and so $\theta\circ\gamma_i(z) = z^{m_i}$ for some $m_i \in Z$ by (3). Hence

$$\theta(z_1,\ldots,z_n) = \theta(\gamma_1(z_1)\cdot\gamma_2(z_2)\cdot\ldots\cdot\gamma_n(z_n))$$

$$= \theta\gamma_1(z_1)\cdot\theta\gamma_2(z_2)\cdot\ldots\cdot\theta\gamma_n(z_n)$$

$$= z_1^{m_1}\cdot z_2^{m_2}\cdot\ldots\cdot z_n^{m_n} \;. \;\; //$$

Theorem:

(1) Every endomorphism $A: K^n \to K^n$ is of the form:

$$A(z_1, \ldots, z_n) = (z_1^{a_{11}} \cdot \ldots \cdot z_n^{a_{1n}}, \ldots, z_1^{a_{n1}} \cdot \ldots \cdot z_n^{a_{nn}})$$

where $a_{ij} \in Z$. In additive notation,

$$A\left(\begin{pmatrix} x_1 \\ \vdots \\ x_n \end{pmatrix} + Z^n\right) = [a_{ij}] \begin{pmatrix} x_1 \\ \vdots \\ x_n \end{pmatrix} + Z^n.$$

(2) A maps K^n onto K^n iff $\det[a_{ij}] \neq 0$.

(3) A is an automorphism of K^n iff $\det[a_{ij}] = \pm 1$.

Proof: (1) Let $\pi_i: K^n \to K$ be the projection to the i-th coordinate. Then $\pi_i \circ A: K^n \to K$ is a homomorphism, so by (4) of the previous theorem

$$\pi_i \circ A(z_1, \ldots, z_n) = z_1^{a_{i1}} \cdot z_2^{a_{i2}} \cdot \ldots \cdot z_n^{a_{in}}$$

where $a_{ij} \in Z$.

(2) Assume $\det[a_{ij}] = 0$. $\exists$ $m_1, \ldots, m_n$ integers not all zero $\ni$ $m_1 A_1 + \ldots + m_n A_n = 0$ where A_i is the i-th row of A. Then each point $(\omega_1, \ldots, \omega_n)$ of K^n in the image of A satisfies $\omega_1^{m_1} \ldots \omega_n^{m_n} = 1$. Thus $A(K^n) \neq K^n$. Conversely if $A(K^n) \neq K^n$ then the points of $A(K^n)$ are annihilated by a nontrivial character of K^n, say $(z_1, \ldots, z_n) \to z_1^{m_1} \cdot \ldots \cdot z_n^{m_n}$ (this is by (5) of §5). Then $m_1 A_1 + \ldots + m_n A_n = 0$ and so $\det[a_{ij}] = 0$.

(3) If A is an automorphism represented by a matrix $[A]$ then A^{-1} is also an automorphism represented by a matrix $[B]$, and since $AA^{-1} = I = A^{-1}A$ we have that $[B] = [A]^{-1}$. Since $[B]$ is an integer matrix, $\det[A] = \pm 1$. Conversely, if $\det[A] = \pm 1$, $[A]^{-1}$ has integer entries and if B is the endomorphism of K^n it defines we have $AB = BA = I$. //

Notation:

If A is an endomorphism of the n-torus, $[A]$ will always denote the associated matrix and $\tilde{A}$ will denote the linear transformation of R^n determined by $[A]$. So if $\pi: R^n \to R^n/Z^n$ is the natural projection ($\pi(x) = x + Z^n$) we have $\pi\tilde{A} = A\pi$.

Let $A: K^n \to K^n$ be an endomorphism. We now consider how the map $\hat{A}: \hat{K}^n \to \hat{K}^n$ (introduced in §5) acts as a map of Z^n when $\hat{K}^n$ is identified with Z^n by the isomorphism:

$$\gamma \mapsto \begin{pmatrix} m_1 \\ m_2 \\ \vdots \\ m_n \end{pmatrix} \quad \text{when} \quad \gamma(z_1, z_2, \ldots, z_n) = z_1^{m_1} \cdot z_2^{m_2} \cdot \ldots \cdot z_n^{m_n} \ .$$

One readily checks that the endomorphism $\hat{A}: Z^n \to Z^n$ is given by

$$\hat{A} \begin{pmatrix} m_1 \\ \vdots \\ m_n \end{pmatrix} = [A]_t \begin{pmatrix} m_1 \\ \vdots \\ m_n \end{pmatrix}$$

where $[A]_t$ denotes the transpose of the matrix $[A]$.

Chapter 1: Measure-Preserving Transformations

§1. Examples

Suppose (X_1, B_1, m_1), (X_2, B_2, m_2) are probability spaces.

Definition 1.1:

a) $T: X_1 \to X_2$ is <u>measurable</u> if $T^{-1}(B_2) \subset B_1$ (i.e., $B_2 \in B_2 \Rightarrow T^{-1}B_2 \in B_1$).

b) $T: X_1 \to X_2$ is <u>measure-preserving</u> if T is measurable and $m_1(T^{-1}(B_2)) = m_2(B_2)$ $\forall B_2 \in B_2$.

c) We say that $T: X_1 \to X_2$ is an <u>invertible</u> <u>measure-preserving</u> <u>transformation</u> if T is measure-preserving, bijective, and T^{-1} is also measure-preserving.

Remarks:

(1) We should write $T: (X_1, B_1, m_1) \to (X_2, B_2, m_2)$ since the measure-preserving property depends on the B's and m's.

(2) If $T: X_1 \to X_2$ and $S: X_2 \to X_3$ are measure-preserving so is $S \circ T: X_1 \to X_3$.

(3) Measure-preserving transformations are the structure preserving maps (morphisms) between measure spaces.

(4) We shall be mainly interested in the case $(X_1, B_1, m_1) = (X_2, B_2, m_2)$ since we wish to study T^n (see §1, Ch. 0).

In practice it would be difficult to check, using Defs. 1.1, whether a given transformation is measure-preserving or not since one usually does not have explicit knowledge of all the members of B. However we often do have explicit knowledge of an algebra A generating B (for example, when X is the unit interval A may be all finite unions of intervals, and when X is a direct product space

A may be the collection of all finite unions of measurable rectangles). The following result is therefore desirable in checking
whether transformations are measure-preserving or not.

Theorem 1.1:

Suppose (X_1,B_1,m_1) , (X_2,B_2,m_2) are probability spaces and
T: $X_1 \to X_2$ is a map. Let A_2 be an algebra which generates B_2.
If $A_2 \in A_2 \Rightarrow T^{-1}(A_2) \in B_1$ and $m_1(T^{-1}(A_2)) = m_2(A_2)$ then T is
measure-preserving.

Proof: Let $C_2 = \{B \in B_2: T^{-1}(B) \in B_1, m_1(T^{-1}(B)) = m_2(B)\}$; we
want to show that $C_2 = B_2$. However $A_2 \subseteq C_2$ and C_2 is easily
seen to be a monotone class, so the result follows since the σ-algebra
generated by C_2 is the monotone class generated by A_2. //

Examples of Measure-Preserving Transformations:

(1) I = identity on (X,B,m) is obviously measure-preserving.

(2) Let $K = \{z \in C: |z| = 1\}$, B = Borel sets, and m = Haar measure.
Define T: K → K by T(z) = az where a is a fixed element of K.
T is measure-preserving since m is Haar measure.

(3) The transformation T(x) = ax defined on any compact group G
(where a is a fixed element of G) preserves Haar measure.

(4) Any continuous endomorphism of a compact group onto itself preserves Haar measure.

Proof: Let A: G → G be a continuous endomorphism, and
m = Haar measure on G. Let $\mu(E) = m(A^{-1}(E))$. μ is a Borel probability measure and

$$\mu(Ax \cdot E) = m(A^{-1}(Ax \cdot E)) = m(x \cdot A^{-1}E) = \mu(E).$$

Since A maps G onto G, μ = m by the uniqueness property of Haar
measure. //

For example, $T(z) = z^n$ preserves Haar measure on the unit circle.

(5) Any affine transformation of a compact group G preserves Haar measure. An <u>affine</u> <u>transformation</u> is a map of the form $T(x) = a \cdot A(x)$ where $a \in G$ is fixed and $A: G \to G$ is a surjective endomorphism. T is measure-preserving because it is the composition of measure-preserving transformations. When $A = I$ we have example (3) and when a is the identity element of G we have example (4).

(6) Let $Y = \{0, \ldots, k-1\}$, and give measure p_i to i such that $\sum_{i=0}^{k-1} p_i = 1$. We let $X = \prod_{-\infty}^{\infty} Y$ together with the direct product measure. Define $T: X \to X$ by:

$$T(\{x_i\}) = \{y_i\}, \quad \text{where } y_i = x_{i+1}.$$

T preserves the measure of each measurable rectangle and thus it preserves the measure of sets which are finite disjoint unions of measurable rectangles. By Theorem 1.1 T is measure-preserving. We call T the <u>two-sided</u> $(p_0, \ldots, p_{k-1})$-<u>shift</u>.

(7) Let Y be as above, $X = \prod_{0}^{\infty} Y$ with the direct product measure. Let $T: X \to X$ be defined by

$$(x_0, x_1, \ldots) \mapsto (x_1, x_2, \ldots).$$

By an analogous argument to the one in example (6) we see that T is measure-preserving. We call T the <u>one-sided</u> $(p_0, \ldots, p_{k-1})$-<u>shift</u>.

(8) Let I^2 be the unit square equipped with Lebesgue measure and I the unit interval with Lebesgue measure. Then $p: I^2 \to I$ defined by $p(x,y) = x$ is measure-preserving.

Given any set X_1, any probability space (X_2, B_2, m_2) and any map $T: X_1 \overset{onto}{\to} X_2$ we can choose a σ-algebra B_1 and a measure m_1 on X_1 to make T measure-preserving. In fact let $B_1 = T^{-1} B_2$ and define m_1 by $m_1(T^{-1} B_2) = m_2(B_2)$.

Conversely, if (X_1, B_1, m_1) is any probability space, X_2 any set

onto
and $T: X_1 \to X_2$ any map, then we can choose a σ-algebra B_2 and a measure m_2 on X_2 so that T is measure-preserving. Put

$$B_2 = \{B: B \subset X_2 \text{ and } T^{-1}B \in B_1\}$$

and $m_2(B) = m_1(T^{-1}B)$ for $B \in B_2$.

§2. Problems in Measure Theoretic Ergodic Theory

(a) External Problems:

How do we apply measure theoretic ergodic theory to other branches of mathematics and physics? In these applications one has a space X with some structure on it, and a map T of X which preserves this structure. To apply the theory of measure preserving transformations one needs an invariant measure for T which acts "nicely" with respect to the structure on X. For example, if X is a topological space we would like the measure to be a Borel measure which is positive on non-empty open sets.

Examples:

(1) Hamiltonian Mechanics: Here one has a one-parameter group of diffeomorphisms of a manifold and there is a smooth measure on the manifold preserved by each diffeomorphism.

(2) Number Theory: To study continued fractions one studies $T: [0,1) \to [0,1)$ given by:

$$T(x) = \begin{cases} 0 & \text{if } x = 0 \\ \{1/x\} & \text{if } x \neq 0 \} \end{cases}$$

where the $\{y\}$ denotes the fractional part of y. T preserves the Gauss measure on $[0,1)$ which is given by:

$$m(A) = \frac{1}{\log 2} \int_A \frac{1}{1+x} \, dx, \quad A \subseteq [0,1).$$

(b)   Internal Problems:

The main internal problem in measure theoretic ergodic theory is: Given two measure preserving transformations when are they isomorphic? (i.e., when can we consider them to be the same?) We look for invariants. A property P is an invariant if when T_1 has the property P and T_2 is isomorphic to T_1 then T_2 has the property P. Invariants give good negative answers, i.e., if T_1 has property P and T_2 does not, then T_1 and T_2 are not isomorphic. The invariants we shall study are of two types:--spectral invariants and entropy.

Before discussing the notion of isomorphism we shall introduce some general concepts such as recurrence and mixing.

§3. Recurrence

Theorem 1.2: (Poincaré Recurrence Theorem)

Let T be a measure-preserving transformation of a probability space $(X, \mathcal{B}, m)$. Let $E \in \mathcal{B}$, $m(E) > 0$. Then almost all points of E return infinitely often to E under positive iteration by T, (in fact, we have the stronger result that: $\exists\ F \subset E$, $m(F) = m(E)$ $\ni$ if $x \in F\ \exists$ integers $0 < n_1 < n_2\ \ldots\ \ni\ T^{n_i}(x) \in F\ \forall\ i$).

Proof: For $N \geq 0$ let $E_N = \bigcup\limits_{n=N}^{\infty} T^{-n}(E)$. We have $T^{-1}(E_N) = E_{N+1}$, $E_N \subset E_{N-1} \subset E_{N-2} \subset \ldots \subset E_0$, $E \subset E_0$, and $m(E_{N+1}) = m(T^{-1}(E_N)) = m(E_N)$ since T is measure-preserving. Therefore, for each N, $m(E_N) = m(E_0)$ and,

$$m(\bigcap\limits_{N=0}^{\infty} E_N) = m(E_0).$$

$\bigcap\limits_{N=0}^{\infty} E_N = \bigcap\limits_{N=0}^{\infty} \bigcup\limits_{n=N}^{\infty} T^{-n}E$, which is the set of all points entering E infinitely often under positive iteration by T.   Moreover

$F = E \cap (\bigcap\limits_{N=0}^{\infty} E_N)$ consists of all points of E which enter E infinitely often under positive iterates of T. Since $\bigcap_{N=0}^{\infty} E_N \subset E_0$

and both sets have the same measure,

$$m(F) = m(E \cap \bigcap_{N=0}^{\infty} E_N) = m(E \cap E_0) = m(E).$$

It remains to show a point of F returns to F infinitely often.
Let $x \in F$, then $\exists\ 0 < n_1 < n_2\ \ldots \ni\ T^{n_i}(x) \in E\ \forall\ i$. Consider
$T^{n_1}(x)$. $T^{n_1}(x) \in E$ and enters E infinitely often under positive
iterates, namely, n_2-n_1, n_3-n_1, $\ldots$ since $T^{n_i-n_1}(T^{n_1}(x))=T^{n_i}(x) \in E$;
thus $T^{n_1}(x) \in F$. Similarly one shows $T^{n_i}(x) \in F\ \forall\ i$. //

Remark:

We do not need that T be measure-preserving in the hypothesis;
we need only assume that T is _incompressible_, i.e., if $B \in B$,
$T^{-1}B \subset B$ then $m(B) = m(T^{-1}B)$.

§4. _Ergodicity_

Let (X,B,m) be a probability space and $T: X \to X$ be a measure-
preserving transformation. If $T^{-1}B = B$ for $B \in B$, then
$T^{-1}(X \setminus B) = X \setminus B$ and we could study T in two separate parts,
namely $T|_B$ and $T|_{X \setminus B}$. If $0 < m(B) < 1$ this has simplified the
study of T. We need a concept of irreducibility for measure-
preserving transformations, such that if T has this irreducibility
property then the study of T cannot be split into two parts as
above. Ergodicity is such a concept. Also, we would like some way
of splitting a measure-preserving transformation into ergodic parts
in a canonical way. This can be done in reasonably well behaved
measure spaces. (See Rohlin [3].)

Definition 1.2:

T: $(X,B,m) \to (X,B,m)$ is _ergodic_ if for $B \in B$, $T^{-1}B = B \Rightarrow m(B) = 0$
or $m(B) = 1$.

Theorem 1.3:

The following are equivalent for measure-preserving T: X → X:

(1) T is ergodic.

(2) $m(T^{-1}B \triangle B) = 0$, $B \in \mathcal{B} \Rightarrow m(B) = 0$ or 1.

(3) ∀ A,B ∈ $\mathcal{B}$, m(A),m(B) > 0 ∃ n > 0 ∍ $m(T^{-n}A \cap B) > 0$.

Proof: (1) ⇒ (2). Suppose $m(T^{-1}B \triangle B) = 0$. Let

$$B_\infty = \bigcap_{n=0}^{\infty} \bigcup_{i=n}^{\infty} T^{-i}B \in \mathcal{B}. \quad \text{Then}$$

$$B_\infty = (B \cup T^{-1}B \cup T^{-2}B \cup \dots) \cap (T^{-1}B \cup T^{-2}B \cup \dots) \cap (T^{-2}B \cup T^{-3}B \cup \dots) \cap \dots;$$

and therefore $T^{-1}B_\infty = B_\infty$ and $m(B_\infty) = m(B)$. Hence, by (1),

$m(B_\infty) = 0$ or 1, and therefore m(B) = 0 or 1.

(2) ⇒ (3). Let m(A) > 0, m(B) > 0 and suppose (3) is false,

i.e., ∀ n > 0 $m(T^{-n}A \cap B) = 0$. Then

$$m\left(\left(\bigcup_{n=1}^{\infty} T^{-n}A \right) \cap B \right) = 0.$$

Let $A' = \bigcup_{n=1}^{\infty} T^{-n}A$. Then $T^{-1}A' \subset A'$ and $m(A') = m(T^{-1}A')$; so that

$m(T^{-1}A' \triangle A') = 0$, and by (2) we have m(A') = 0 or 1. But $T^{-1}A \subset A'$

and T is measure-preserving, so that m(A') = 1. But this contra-

dicts the above fact that $m(A' \cap B) = 0$.

(3) ⇒ (1). Suppose (1) is false, i.e., ∃ B ∈ $\mathcal{B}$, $T^{-1}B = B$ and

0 < m(B) < 1. Then $m(T^{-n}B \cap (X \backslash B)) = 0$ ∀ n > 0 which contradicts

(3). //

A characterization of ergodicity in terms of functions is given

by the following results.

Theorem 1.4:

Let $T: (X,\mathcal{B},m) \to (X,\mathcal{B},m)$ be measure-preserving; then the following are equivalent:

(1) T is ergodic.

(2) Whenever f is measurable and $(f \circ T)(x) = f(x)$ $\forall$ $x \in X$ then f is constant a.e.

(3) Whenever f is measurable and $(f \circ T)(x) = f(x)$ a.e. then f is constant a.e.

(4) Whenever $f \in L^2(m)$ and $(f \circ T)(x) = f(x)$ $\forall$ $x \in X$ then f is constant a.e.

(5) Whenever $f \in L^2(m)$ and $(f \circ T)(x) = f(x)$ a.e. then f is constant a.e.

Proof: (1) $\Rightarrow$ (3). Suppose f is measurable and $f \circ T = f$ a.e.; while T is ergodic. We can assume that f is real-valued for if f is complex-valued we can consider the real and imaginary parts separately. Define

$$X(k,n) = \{x: k/2^n \leq f(x) < (k+1)/2^n\} \quad k \in \mathbb{Z}, \quad n > 0.$$

We have

$$T^{-1}X(k,n) \triangle X(k,n) \subset \{x: (f \circ T)(x) \neq f(x)\}$$

and hence $m(T^{-1}X(k,n) \triangle X(k,n)) = 0$ so that by (2) of Theorem 1.3 $m(X(k,n)) = 0$ or 1.

Fix n, then $\bigcup_{k \in \mathbb{Z}} X(k,n) = X$ which is a disjoint union; so for each n $\exists$ unique k_n $\ni$ $m(X(k_n,n)) = 1$. Let $Y = \bigcap_{n=1}^{\infty} X(k_n,n)$. Then $m(Y) = 1$ and f is constant on Y so that f is constant a.e.

Trivially we have (3) $\Rightarrow$ (2) $\Rightarrow$ (4), (5) $\Rightarrow$ (4), and (3) $\Rightarrow$ (5). So it remains to show:

(4) $\Rightarrow$ (1). Suppose $T^{-1}E = E$, $E \in \mathcal{B}$. Then $\chi_E \in L^2(m)$ and

$(\chi_E \circ T)(x) = \chi_E(x)$ $\forall$ $x \in X$ so, by (4) χ_E is constant a.e. Hence $m(E) = \int \chi_E dm = 0$ or 1. //

Note:

A similar characterization in terms of $L^1(m)$ functions is true, since in the last part of the proof χ_E is in $L^1(m)$ as well as $L^2(m)$. Also we could use real $L^1(m)$ or $L^2(m)$ spaces.

Remark:

The following remark comes later (Theorem 5.5) but we preview it now in order to analyze our examples.

Let X be a compact metric space and m a Borel probability measure on X which gives positive measure to every non-empty open set. If $T: X \to X$ is continuous and ergodic with respect to m then $m(\{x \mid \{T^n x \mid n \geq 0 \text{ is dense}\}\}) = 1$.

Proof: Let $\{U_n\}_{n=1}^{\infty}$ be a base for the topology of X. $\{T^n x \mid n \geq 0\}$ is dense in X $\Leftrightarrow$ $x \in \bigcap_{n=1}^{\infty} \bigcup_{k=0}^{\infty} T^{-k} U_n$. Since $T^{-1}(\bigcup_{k=0}^{\infty} T^{-k} U_n) \subset \bigcup_{k=0}^{\infty} T^{-k} U_n$ and T is measure-preserving and ergodic we have $m(\bigcup_{k=0}^{\infty} T^{-k} U_n) = 0$ or 1. Since $\bigcup_{k=0}^{\infty} T^{-k} U_n$ is a non-empty open set we have $m(\bigcup_{k=0}^{\infty} T^{-k} U_n) = 1$. The result follows. //

Note that this result is applicable when m is Haar measure on a compact metric group and T is an affine transformation.

Examples:

We shall now see when the examples of §1 are ergodic.

(1) I on $(X, \mathcal{B}, m)$ is ergodic iff all members of $\mathcal{B}$ have measure 0 or 1.

(2) Consider $T: K \to K$, $T(z) = az$. T is ergodic iff a is not a root of unity.

Proof: Suppose a is a root of unity, then $a^p = 1$ for some

$p \neq 0$. Let $f(z) = z^p$; then clearly $f \neq$ constant a.e. and $f \circ T = f$.
Conversely, suppose a is not a root of unity and $f \circ T = f$, $f \in L^2(m)$.
Let $f(z) = \sum_{n=-\infty}^{\infty} b_n z^n$ be its Fourier series. Then by above,
$\sum b_n a^n z^n = \sum b_n z^n$ and therefore $b_n(a^n - 1) = 0$. If $n \neq 0$ then
$b_n = 0$, and so f is constant a.e. //

(3) Let $T(x) = ax$ on a compact metric group G, then T is ergodic iff $\{a^n\}_{n \in Z}$ is dense in G. In particular, T ergodic $\Rightarrow$
G is abelian.

Proof: Suppose firstly that $Tx = ax$ is ergodic. By the above
remark it follows that $\{a^n x_0\}_0^{\infty}$ is dense for some x_0 and so $\{a^n\}_0^{\infty}$
is dense since if $y \in G \ \exists \ \{n_i\}$ with $a^{n_i} x_0 \to y x_0$ i.e., $a^{n_i} \to y$.
Conversely, suppose $\{a^n\}_{n \in Z}$ is dense in G. This implies G is
abelian. Let $f \in L^2(m)$ and $f \circ T = f$. By (7) of §5 of Chapter 0
f can be represented as $\sum_i b_i \gamma_i$, where $\gamma_i \in \hat{G}$. Then
$\sum_i b_i \gamma_i(a) \gamma_i(x) = \sum_i b_i \gamma_i(x)$ so that if $b_i \neq 0$ then $\gamma_i(a) = 1$
and so $\gamma_i \equiv 1$. Therefore only the constant term of the Fourier
series of f can be non-zero, i.e., f is constant a.e. //

(4) For an endomorphism A of a compact metric group G necessary
and sufficient conditions for ergodicity are known. When G is
abelian, Halmos [4] and Rohlin independently proved that A is
ergodic $\Leftrightarrow$ whenever $\gamma \circ A^n = \gamma$, $n > 0$, then $\gamma = 1$. Before proving
the general result we will illustrate the proof by showing that
$A(z) = z^2$ on K is ergodic.

Suppose $f \circ A = f$, $f \in L^2(m)$. We have that if $f(z) = \sum_{n=-\infty}^{\infty} a_n z^n$
then $\sum a_n z^{2n} = \sum a_n z^n$ and therefore $a_n = a_{2n} = a_{4n} = \cdots$. So
if $n \neq 0$ we must have $a_n = 0$ because $\sum_{-\infty}^{\infty} |a_n|^2 < \infty$. Therefore
f is constant a.e.

Proof of the general result: Suppose that whenever $\gamma A^n = \gamma$ we have $\gamma = 1$, and let $f \circ A = f$ with $f \in L^2(m)$. Let $f(x)$ have the Fourier series $\sum a_n \gamma_n$ where $\gamma_n \in \hat{G}$ and $\sum |a_n|^2 < \infty$. Then $\sum a_n \gamma_n(Ax) = \sum a_n \gamma_n(x)$, so that if γ_n, $\gamma_n \circ A$, $\gamma_n \circ A^2$, ... are all distinct their corresponding coefficients are equal and therefore zero. So if $a_n \neq 0$, $\gamma_n(A^p) = \gamma_n$ for some $p > 0$. Then $\gamma_n = 1$ by assumption and so f is constant a.e.

Conversely let A be ergodic and $\gamma A^n = \gamma$, $n > 0$. If n is the least such integer, $f = \gamma + \gamma A + ... + \gamma A^{n-1}$ is invariant under A and not a.e. constant (being the sum of orthogonal functions), contradicting the ergodicity. //

Consider now the case when G is the n-torus K^n. The ergodicity condition then becomes: $A: K^n \to K^n$ is ergodic $\Leftrightarrow$ the matrix $[A]$ has no roots of unity as eigenvalues.

Proof: Recall that, under the identification $\hat{K}^n \simeq Z^n$, if
$$\underline{m} = \begin{pmatrix} m_1 \\ \vdots \\ m_n \end{pmatrix} \in Z^n$$
then $[A]_t \underline{m} = \hat{A}(\underline{m})$. If A is not ergodic then $[A]_t^k \underline{m} = \underline{m}$ for some $\underline{m} \neq \underline{0}$ and $k > 0$. Then $[A]_t^k$ has 1 as an eigenvalue so $[A]_t$ has a k-th root of unity as an eigenvalue.

Conversely if $[A]_t$ has a k-th root of unity as an eigenvalue then $[A]_t^k$ has 1 as an eigenvalue so that $[A]_t^k - I$ induces a singular linear transformation R^n. Hence $[A]_t^k - I$ induces a many-to-one map of Z^n into Z^n and so there exists $\underline{0} \neq \underline{m} \in Z^n$ with $[A]_t^k \underline{m} = \underline{m}$. //

So, for example, all the endomorphisms $z \to z^k$, $|k| > 1$, of K are ergodic.

Chu [1] has considered the case when G is nonabelian. He has shown that a continuous endomorphism of a compact group G onto itself is ergodic if and only if the induced map on the representation

ring R(G) has no finite orbit except the constant functions. The
representation ring, R(G), is the ring generated by the coefficients
of all irreducible unitary representations of G over the complex
field.

(5) For affine transformations of compact metric groups necessary and
sufficient conditions for ergodicity are known. The simplest case is
when G is a compact, connected, metric, abelian group. If Tx =
a·A(x) is an affine transformation of the compact, connected, metric,
abelian group G then the following are equivalent:

(a) T is ergodic.

(b) (i) Whenever $\gamma \circ A^k = \gamma$ for k > 0 then $\gamma \circ A = \gamma$, and,

   (ii) the smallest closed subgroup containing a and BG (where
       $Bx = x^{-1} \cdot A(x)$) is G (i.e., [a,BG] = G).

(c) $\exists$ $x_0 \in G$ with $\{T^n(x_0): n \geq 0\}$ dense in G.

(d) $m(\{x: \{T^n x: n \geq 0 \text{ is dense}\}\}) = 1$.

(Note that conditions (i) and (ii) reduce to the conditions given in
(3) and (4) in the special cases. The equivalence of (a) and (b) was
investigated by Hahn, Hoare and Parry.)

Proof: First note that B is an endomorphism of G and commutes with A.

(b) $\Rightarrow$ (a). Suppose (i) and (ii) of (b) hold. If foT = f,
$f \in L^2(m)$ let $f = \sum b_i \gamma_i$, $\gamma_i \in \hat{G}$. Then

$$\sum_i b_i \gamma_i(a) \gamma_i(Ax) = \sum_i b_i \gamma_i(x). \qquad (*)$$

So if γ_i, $\gamma_i \circ A$, $\gamma_i \circ A^2$, ... are all distinct then $b_i = 0$ or else
$\sum |b_i|^2 < \infty$ is violated. Hence, if $b_i \neq 0$ then $\gamma_i \circ A^n = \gamma_i$ for
some n > 0, and by (i) $\gamma \circ A = \gamma$. But then (*) implies $\gamma_i(a) = 1$
and so $\gamma_i(x) = 1 \ \forall \ x \in [a,BG]$ and by (ii) $\gamma_i = 1$. So f is constant a.e.

(a) $\Rightarrow$ (d). This follows by the remark above.

(d) $\Rightarrow$ (c) is trivial.

(c) $\Rightarrow$ (b). It remains to show that if $\exists$ $x_0 \in G$ with $\{T^n x_0 : n \geq 0\}$ dense in G then conditions (i) and (ii) of (b) hold. Suppose $\gamma \circ A^k = \gamma$, $k \geq 1$, $\gamma \in \hat{G}$. Let $\gamma_1 = \gamma \circ B$. Then $\gamma_1(T^k x) = \gamma_1(a \cdot Aa \cdot \ldots \cdot A^{k-1}a)\gamma_1(A^k x) = \gamma(a^{-1}A^k a)\gamma_1(x) = \gamma_1(x)$. Hence γ_1 assumes only the finite number of values $\gamma_1(x_0)$, $\gamma_1(Tx_0)$, ..., $\gamma_1(T^{k-1}x_0)$ on the dense set $\{T^n x_0 : n \geq 0\}$ and hence assumes only these values on G. Since G is connected γ_1 must be constant, i.e., $\gamma_1 = 1$. Hence $\gamma A = \gamma$ and condition (i) holds.

If $[a, BG] \neq G \ni \gamma \neq 1$, $\gamma \in \hat{G}$, with $\gamma(a) = 1$, and $\gamma(Bx) = 1$. Then $\gamma(Tx) = \gamma(x)$ and so γ assumes only the value $\gamma(x_0)$ on the dense set $\{T^n x_0 : n \geq 0\}$ and therefore γ is a constant. Hence $\gamma \equiv 1$, a contradiction, and we have shown that condition (c) implies (ii). //

When G is K^n the equivalence of (a) and (b) becomes: $T = a \cdot A$ is ergodic iff

      (i)   the matrix $[A]$ has no proper roots of unity (i.e.,

           other than 1) as eigenvalues,

and   (ii) $[a, BK^n] = K^n$.

This is easily proved by a method similar to the one used in (4) for the endomorphism case.

Conditions for ergodicity of affine transformations of compact nonabelian groups may also be found in Chu [1].

(6) The 2-sided $(p_0, \ldots, p_{k-1})$-shift is ergodic.

Proof: Let A = the algebra generated by finite unions of measurable rectangles. Suppose $T^{-1}E = E$, $E \in B$. Let $\varepsilon > 0$ be given, and choose $A \in A \ni m(E \Delta A) < \varepsilon$; thus

$$|m(E) - m(A)| = |m(E \cap A) + m(E \backslash A) - m(A \cap E) - m(A \backslash E)|$$

$$< m(E \backslash A) + m(A \backslash E) < \varepsilon.$$

Choose n so large that $B = T^{-n}A$ depends upon different coordinates from A; so, $m(B \cap A) = m(B)m(A) = m(A)^2$.

$$m(E \Delta B) = m(T^{-n}E \Delta T^{-n}A) = m(E \Delta A) < \varepsilon$$

and since $E \Delta (A \cap B) \subset E \Delta A \cup E \Delta B$ we have $m(E \Delta (A \cap B)) < 2\varepsilon$, hence

$$|m(E) - m(A \cap B)| < 2\varepsilon$$

and

$$|m(E) - m(E)^2| \le |m(E) - m(A \cap B)| + |m(A \cap B) - m(E)^2|$$

$$< 2\varepsilon + |m(A)^2 - m(E)^2|$$

$$\le 2\varepsilon + m(A)|m(A) - m(E)| + m(E)|m(A) - m(E)|$$

$$< 4\varepsilon$$

since $m(A), m(E) \le 1$. Since ε is arbitrary $m(E) = m(E)^2$ which implies that $m(E) = 0$ or 1. //

(7) By a similar argument, we see that the 1-sided $(p_0, \ldots, p_{k-1})$-shift is ergodic.

§5. The Ergodic Theorem

The first major result in ergodic theory was proved in 1931 by G. D. Birkhoff [1].

Theorem 1.5: (Birkhoff Ergodic Theorem)

Suppose $T: (X, B, m) \to (X, B, m)$ is measure-preserving (where we allow (X, B, m) to be σ-finite) and $f \in L^1(m)$. Then $\frac{1}{n} \sum_{i=0}^{n-1} f(T^i(x))$ converges a.e. to a function $f^* \in L^1(m)$. Also, $f^* \circ T = f^*$ a.e., and if $m(X) < \infty$, $\int f^* dm = \int f dm$.

Note:

If T is ergodic then $f^* = $ a constant a.e. and if $m(X) < \infty$

then $f^* = \dfrac{1}{m(X)} \displaystyle\int f\, dm$ a.e.

Motivation:

(i) Suppose $T: (X,B,m) \to (X,B,m)$ is measure-preserving and $E \in B$.
For $x \in X$, we could ask with what frequency do the elements of the
set $\{x, T(x), T^2(x), \ldots\}$ lie in the set E?

Clearly $T^i(x) \in E$ iff $\chi_E T^i(x) = 1$, so the number of elements
of $\{x, T(x), \ldots, T^{n-1}(x)\}$ in E is $\displaystyle\sum_{k=0}^{n-1} \chi_E T^k(x)$; and so the relative number of elements of $\{x, T(x), \ldots, T^{n-1}(x)\}$ in E equals
$\dfrac{1}{n}\displaystyle\sum_{i=0}^{n-1} \chi_E T^k(x)$. If $m(x) = 1$ and T is ergodic then $\dfrac{1}{n}\displaystyle\sum_{i=0}^{n-1} \chi_E T^i(x) \to$
$m(E)$ a.e. by the note; and thus the orbit of almost every point of X
enters the set E with asymptotic relative frequency $m(E)$.

(ii) We define the time mean of f to be

$$\lim_{n\to\infty} \frac{1}{n}\sum_{i=0}^{n-1} f(T^i(x))$$

and the phase or space mean of f to be

$$\frac{1}{m(X)}\int_X f(x)dm.$$

The ergodic theorem implies these means are equal if T is ergodic.
(The converse is also true.) So, it is important to verify ergodicity
for transformations arising in physics. This application to time
means and space means is more realistic in the case of a 1-parameter
flow $\{T_t\}$ of measure-preserving transformations. The ergodic theo-
rem then asserts $\lim\limits_{T\to\infty} \dfrac{1}{T}\displaystyle\int_0^T f(T_t x)dt$ exists a.e. for $f \in L^1(m)$ and
equals $\dfrac{1}{m(X)}\displaystyle\int_X f\, dm$ in the ergodic case if the map $(t,x) \to T_t x \cdot$is

measurable.

An Application to Number Theory

Borel's Theorem on Normal Numbers:

Almost all numbers in [0,1) are normal to base 2, i.e.,

$$\frac{1}{n} \cdot \left(\begin{array}{l} \text{the number of 1's in the first } n \text{ digits} \\ \text{of the binary expansion of } x \in [0,1) \end{array} \right) \rightarrow \frac{1}{2} \text{ a.e.}$$

Proof: Let $T: [0,1) \rightarrow [0,1)$ be defined by $T(x) = 2x \bmod 1$. We know that T preserves Lebesgue measure and is ergodic, by example 4 at the end of §4.

Suppose $x = \dfrac{a_1}{2} + \dfrac{a_2}{2^2} + \ldots$ has a unique binary expansion. Then

$$T(x) = T\left(\frac{a_1}{2} + \frac{a_2}{2^2} + \frac{a_3}{2^3} + \ldots \right) = \frac{a_2}{2} + \frac{a_3}{2^2} + \ldots .$$ Let $f(x) = \chi_{[\frac{1}{2},1)}(x)$. Then

$$f(T^i(x)) = f\left(\frac{a_{i+1}}{2} + \frac{a_{i+2}}{2^2} + \ldots \right) = \begin{cases} 1 & \text{iff } a_{i+1} = 1 \\ \\ 0 & \text{iff } a_{i+1} = 0 \end{cases} .$$

Hence, the number of 1's in the first n digits of the dyadic expansion of x is $\sum_{i=0}^{n-1} f(T^i(x))$. Dividing both sides of this equality by n and applying the ergodic theorem we see that

$$\frac{1}{n} \sum_{i=0}^{n-1} f(T^i x) \xrightarrow[a.e.]{} \int \chi_{[\frac{1}{2},1)} dm = \frac{1}{2} \text{ a.e.}$$

(using the fact that the binary-rational points form a set of Lebesgue measure zero). //

The ergodic theorem can be applied to give other number theoretic results. Some are obtained in Billingsley [1] and Avez-Arnold [1]. We now consider some preliminaries to the proof of the ergodic theorem.

Definition 1.3:

Let $T: (X,\mathcal{B},m) \to (X,\mathcal{B},m)$ be measure-preserving. Define an operator U_T on complex-valued functions on X by:

$$(U_T f)(x) = f(T(x)).$$

We have $U_T L^p(m) \subset L^p(m)$ and, since T is measure-preserving $\|U_T f\|_p = \|f\|_p$. Let $L^p_R(m)$ denote the real-valued $L^p(m)$ functions, then $U_T L^p_R(m) \subset L^p_R(m)$.

To prove Birkhoff's theorem we need:

Theorem 1.6: (<u>Maximal</u> <u>Ergodic</u> <u>Theorem</u>)

Let $U: L^1_R(m) \to L^1_R(m)$ be a positive linear operator (i.e., $f \geq 0 \Rightarrow Uf \geq 0$) which has norm ≤ 1. Let $N > 0$ be an integer. Define $f_0 = 0$, $f_n = f + Uf + U^2 f + \ldots + U^{n-1} f$, and $F_N = \max_{0 \leq n \leq N} f_n \geq 0$. Then $\int_{\{x : F_N(x) > 0\}} f\, dm \geq 0$.

<u>Proof</u>: (due to A. Garsia) Clearly $F_N \in L^1_R(m)$. We have for $0 \leq n \leq N$ $F_N \geq f_n$ so, $UF_N \geq Uf_n$ by positivity, and hence $UF_N + f \geq f_{n+1}$. Therefore

$$UF_N(x) + f(x) \geq \max_{1 \leq n \leq N} f_n(x)$$

$$= \max_{0 \leq n \leq N} f_n(x) \quad \text{when} \quad F_N(x) > 0$$

$$= F_N(x).$$

Thus $f \geq F_N - UF_N$ on $A = \{x : F_N(x) > 0\}$, so

$$\int_A f \, dm \geq \int_A F_N \, dm - \int_A U F_N \, dm$$

$$= \int_X F_N \, dm - \int_A U F_N \, dm \quad \text{since } F_N = 0 \text{ on } X \backslash A.$$

$$\geq \int_X F_N \, dm - \int_X U F_N \, dm \quad \text{since } F_N \geq 0 \Rightarrow U F_N \geq 0.$$

$$\geq 0 \quad \text{since } \|U\| \leq 1. \quad //$$

Remark:

The conditions of Theorem 1.6 hold if $U = U_T$ for measure-preserving T.

Corollary 1.6:

Let $T: X \to X$ be measure-preserving. If $g \in L_R^1(m)$ and

$$B_\alpha = \{x \in X: \sup_{n \geq 1} \frac{1}{n} \sum_{m=0}^{n-1} g(T^m(x)) > \alpha\}$$

then

$$\int_{B_\alpha \cap A} g \, dm \geq \alpha m(B_\alpha \cap A)$$

if $T^{-1}A = A$ and $m(A) < \infty$.

Proof: We first prove this result under the assumptions $m(X) < \infty$ and $A = X$. Let $f = g - \alpha$, then $B_\alpha = \bigcup_{N=0}^{\infty} \{x: F_N(x) > 0\}$ so that $\int_{B_\alpha} f \, dm > 0$ by Theorem 1.6 and therefore $\int_{B_\alpha} g \, dm \geq \alpha m(B_\alpha)$. In the general case, using $T|_A$ in the place of T we see that $\int_{A \cap B_\alpha} g \, dm \geq \alpha m(A \cap B_\alpha)$. //

Proof of Birkhoff's Theorem: It suffices to prove the theorem for $f \in L_R^1(m)$. Let $f^*(x) = \overline{\lim_n} \frac{1}{n} \sum_{i=0}^{n-1} f(T^i(x))$ and $f_*(x) = \underline{\lim_n} \frac{1}{n} \sum_{i=0}^{n-1} f(T^i(x))$. We have $f^* \circ T = f^*$, $f_* \circ T = f_*$ because if

34

$a_n(x) = \frac{1}{n} \sum_{i=0}^{n-1} f(T^i x)$ then $\left(\frac{n+1}{n}\right) a_{n+1}(x) - a_n(Tx) = \frac{f(x)}{n}$. For real

numbers $\beta < \alpha$, let

$$E_{\alpha,\beta} = \{x \in X : f_*(x) < \beta, \ \alpha < f^*(x)\}.$$

Then $T^{-1}E_{\alpha,\beta} = E_{\alpha,\beta}$ and

$$E_{\alpha,\beta} \cap \{x \in X : \sup_{n \geq 1} \frac{1}{n} \sum_{i=0}^{n-1} f(T^i(x)) > \alpha\} = E_{\alpha,\beta}.$$

We now prove that $m(E_{\alpha,\beta}) < \infty$ so that we can apply Corollary 1.6.

Suppose $\alpha > 0$. Let $C \subset E_{\alpha,\beta}$ with $m(C) < \infty$. Then $h = f - \alpha\chi_C$
is integrable and by the maximal ergodic theorem

$\int_{\bigcup_{N=0}^{\infty} \{x:H_N(x)>0\}} (f - \alpha\chi_C)dm \geq 0.$ (H_N defined analogously to the F_N in the

maximal ergodic theorem.) But $C \subset \bigcup_{N=0}^{\infty} \{x : H_N(x) > 0\}$ so that

$\int_X |f|dm \geq \alpha m(C)$. Therefore $m(C) \leq \frac{1}{\alpha} \int_X |f|dm$ for every subset

of $E_{\alpha,\beta}$ with finite measure and hence $m(E_{\alpha,\beta}) < \infty$. If $\alpha < 0$ then
$\beta < 0$ so we can apply the above with $-f$ and $-\beta$ replacing f and
α to get $m(E_{\alpha,\beta}) < \infty$.

Let $B_\alpha = \{x \in X : \sup_{n \geq 1} \frac{1}{n} \sum_{i=0}^{n-1} f(T^i x) > \alpha\}$. Then by Corollary 1.6:

$\int_{E_{\alpha,\beta}} f\,dm = \int_{E_{\alpha,\beta} \cap B_\alpha} f\,dm \geq \alpha m(E_{\alpha,\beta} \cap B_\alpha) = \alpha m(E_{\alpha,\beta})$, i.e.,

$$\int_{E_{\alpha,\beta}} f\,dm \geq \alpha m(E_{\alpha,\beta}) . \tag{*}$$

If we replace f, α, β by $-f, -\beta, -\alpha$ respectively we get that
$(-f)^* = -f_*$, $(-f)_* = -f^*$ and

$$\int_{E_{\alpha,\beta}} f\,dm \leq \beta m(E_{\alpha,\beta}). \tag{**}$$

So, if $\alpha > \beta$ then $m(E_{\alpha,\beta}) = 0$, and since

$$\{x: f_*(x) < f^*(x)\} \subseteq \bigcup_{\substack{\beta < \alpha \\ \alpha,\beta \text{ rational}}} E_{\alpha,\beta} ,$$

we have $m\{x: f_*(x) < f^*(x)\} = 0$ i.e., $f^*(x) = f_*(x)$ a.e. Therefore $\frac{1}{n} \sum_{i=0}^{n-1} f(T^i(x))$ converges a.e.

To show $f^* \in L^1(m)$ we use the part of Fatou's Lemma that says for non-negative integrable functions g_n $\underline{\lim} \int g_n dm < \infty$ implies $\underline{\lim}\, g_n$ is integrable. Let

$$g_n(x) = \left| \frac{1}{n} \sum_{i=0}^{n-1} f(T^i(x)) \right| .$$

Then

$$\int g_n dm = \int \left| \frac{1}{n} \sum_{i=0}^{n-1} f(T^i(x)) \right| dm \leq \int |f| dm$$

so that $\underline{\lim} \int g_n dm < \infty$, and by Fatou's Lemma $\underline{\lim}\, g_n = |f_*|$ is integrable. Hence f_* is integrable.

It remains to show that $\int f\, dm = \int f^* dm$ if $m(X) < \infty$. Let $D_k^n = \{x \in X: \frac{k}{n} \leq f^*(x) < \frac{k+1}{n}\}$ where $k \in Z$, $n \geq 1$. For each small $\varepsilon > 0$ we have $D_k^n \cap B_{(\frac{k}{n}-\varepsilon)} = D_k^n$ and by Corollary 1.6 $\int_{D_k^n} f\, dm \geq (\frac{k}{n} - \varepsilon)m(D_k^n)$ so that

$$\int_{D_k^n} f\, dm \geq \frac{k}{n} m(D_k^n). \qquad (\text{***})$$

Then

$$\int_{D_k^n} f^* dm \leq \frac{k+1}{n} m(D_k^n) \leq \frac{1}{n} m(D_k^n) + \int_{D_k^n} f\, dm \quad (\text{by } (\text{***})).$$

Summing over k we get that

$$\int_X f^* dm \ \leq \ \frac{m(X)}{n} + \int_X f \, dm \qquad \forall \ n \geq 1;$$

thus $\int_X f^* dm \leq \int_X f \, dm$ since $m(X) < \infty$. Applying this to $-f$ instead of f gives $\int_X (-f)^* dm \leq \int_X - f \, dm$ i.e., $-\int_X f_* dm \leq -\int_X f \, dm$. Since $f_* = f^*$ a.e. we get that $\int_X f^* dm \geq \int_X f \, dm$.

Hence, $\int f^* dm = \int f \, dm$. //

Corollaries 1.5:

(i) Let $(X, \mathcal{B}, m)$ be a probability space and $T: X \to X$ measure-preserving, then T is ergodic iff $\forall \ A, B \in \mathcal{B}$

$$\frac{1}{n} \sum_{i=0}^{n-1} m(T^{-i}A \cap B) \ \to \ m(A)m(B).$$

Proof: ($\Rightarrow$) Suppose T is ergodic. Putting $f = \chi_A$ in Theorem 1.5 gives $\frac{1}{n} \sum_{i=0}^{n-1} \chi_A(T^i(x)) \to m(A)$ a.e. Multiplying by χ_B:

$$\frac{1}{n} \sum_{i=0}^{n-1} \chi_A(T^i(x)) \chi_B \ \to \ m(A) \chi_B \quad \text{a.e.}$$

By the dominated convergence theorem if we integrate we get

$$\frac{1}{n} \sum_{i=0}^{n-1} m(T^{-i}A \cap B) \ \to \ m(A)m(B).$$

($\Leftarrow$) Let $T^{-1}E = E$, $E \in \mathcal{B}$. Let $A = B = E$. Then $\frac{1}{n} \sum_{i=0}^{n-1} m(E) \to m(E)^2$ so $m(E) = m(E)^2$, hence $= 0$ or 1. //

(ii) L^p Ergodic Theorem: (Von Neumann [1], [2])

Let $1 \leq p < \infty$. Let T be measure-preserving on the probability space $(X, \mathcal{B}, m)$. If $f \in L^p(m)$, $\exists \ f^* \in L^p(m) \ni f^* \circ T = f^*$ a.e. and $\| \frac{1}{n} \sum_{i=0}^{n-1} f(T^i x) - f^*(x) \|_p \ \to \ 0$.

Proof: If g is bounded and measurable then $g \in L^p$ $\forall$ p and

by the ergodic theorem we have that $\frac{1}{n} \sum_{i=0}^{n-1} g(T^i x) \to g^*(x)$ a.e.

Clearly $g^* \in L^\infty(m)$ and hence $g^* \in L^p(m)$. Also,

$|\frac{1}{n} \sum_{i=0}^{n-1} g(T^i x) - g^*(x)|^p \to 0$ a.e. and by the bounded convergence

theorem, $\|\frac{1}{n} \sum_{i=0}^{n-1} g(T^i x) - g^*(x)\|_p \to 0$ i.e., $\forall \ \varepsilon > 0 \ \exists \ N(\varepsilon,g) \ \ni$

if $n > N(\varepsilon,g)$ and $k > 0$

$$\|\frac{1}{n} \sum_{i=0}^{n-1} g(T^i x) - \frac{1}{n+k} \sum_{i=0}^{n+k-1} g(T^i x)\|_p < \varepsilon .$$

Let $f \in L^p(m)$, and $M_n(f)(x) = \frac{1}{n} \sum_{i=0}^{n-1} f(T^i x)$. We must show that

$\{M_n(f)\}$ is a Cauchy sequence in $L^p(m)$. Note that $\|M_n(f)\|_p \leq \|f\|_p$.

Choose $g \in L^\infty(m) \ni \|f - g\|_p < \varepsilon/4$; then

$$\|M_n f - M_{n+k} f\|_p \leq \|M_n f - M_n g\|_p + \|M_n g - M_{n+k} g\|_p + \|M_{n+k} g - M_{n+k} f\|_p$$

$$\leq \varepsilon/4 + \varepsilon/2 + \varepsilon/4 = \varepsilon$$

if $n > N(\varepsilon/2,g)$ and $k > 0$. We have $f^* \circ T = f^*$ a.e. because

$$(\frac{n+1}{n})(M_{n+1} f)(x) - (M_n f)(Tx) = \frac{f(x)}{n} .$$

§6. Mixing

We have seen that T is ergodic iff $\forall \ A,B \in B$,

$$\frac{1}{N} \sum_{i=0}^{N-1} m(T^{-i} A \cap B) \to m(A)m(B).$$

Definitions 1.4:

(i) T is **weak-mixing** if $\forall \ A,B \in B$

$$\frac{1}{N} \sum_{i=0}^{N-1} |m(T^{-i} A \cap B) - m(A)m(B)| \to 0.$$

(ii)   T is <u>strong-mixing</u> if $\forall$ A,B $\in$ B

$$m(T^{-N}A \cap B) \;\rightarrow\; m(A)m(B).$$

<u>Note</u>:

(i)   T strong-mixing $\Rightarrow$ T weak-mixing.

(ii)   T weak-mixing $\Rightarrow$ T ergodic.

This is so because if $\{a_n\}$ is a sequence of real numbers then

$$a_n \;\rightarrow\; 0 \;\Rightarrow\; \frac{1}{n}\sum_{i=0}^{n-1}|a_i| \;\rightarrow\; 0 \;\Rightarrow\; \frac{1}{n}\sum_{i=0}^{n-1}a_i \;\rightarrow\; 0.$$

(Put $a_n = m(T^{-n}A \cap B) - m(A)m(B)$.)

(iii)   An example of an ergodic T which is not weak-mixing is given by T(z) = az on K, where a is not a root of unity. (See the end of this section for the proof.)

(iv)   There are examples of weak-mixing T which are not strong-mixing. Kakutani has an example constructed by combinatorial methods, and Maruyama constructed an example using Gaussian processes. Chacon and Katok-Stepin also have examples. Indeed, if (X,B,m) is a probability space, let $\tau(X)$ denote the collection of all invertible measure-preserving transformations of (X,B,m). If we topologize $\tau(X)$ with the "weak" topology (see Halmos [2]), the class of weak-mixing transformations is of second category while the class of strong mixing transformations is of first category.

The following result shows it suffices to check the convergence properties on an algebra generating B.

<u>Theorem</u> 1.7:

If T: X $\rightarrow$ X is measure-preserving and A is an algebra generating B then

(i) T is ergodic iff $\forall$ A,B $\in$ A

$$\frac{1}{n} \sum_{i=0}^{n-1} m(T^{-i}A \cap B) \rightarrow m(A)m(B),$$

(ii) T is weak-mixing iff $\forall$ A,B $\in$ A

$$\frac{1}{n} \sum_{i=0}^{n-1} |m(T^{-i}A \cap B) - m(A)m(B)| \rightarrow 0, \qquad \text{and}$$

(iii) T is strong-mixing iff $\forall$ A,B $\in$ A

$$m(T^{-n}A \cap B) \rightarrow m(A)m(B).$$

Proof: Let $\varepsilon > 0$ and E,F $\in$ B. Choose $E_0, F_0 \in$ A with $m(E \Delta E_0) < \varepsilon$, $m(F \Delta F_0) < \varepsilon$. Then

$$m((T^{-n}E \cap F) \Delta (T^{-n}E_0 \cap F_0)) < 2\varepsilon$$

and therefore

$$|m(T^{-n}E \cap F) - m(T^{-n}E_0 \cap F_0)| < 2\varepsilon.$$

Hence

$$|\frac{1}{n} \sum_{k=0}^{n-1} m(T^{-k}E \cap F - m(E)m(F)|$$

$$\leq |\frac{1}{n} \sum_{k=0}^{n-1} [m(T^{-k}E \cap F) - m(T^{-k}E_0 \cap F_0)]|$$

$$+ |\frac{1}{n} \sum_{k=0}^{n-1} m(T^{-k}E_0 \cap F_0) - m(E_0)m(F_0)|$$

$$+ |\frac{1}{n} \sum_{k=0}^{n-1} m(E_0)m(F_0) - m(E_0)m(F)|$$

$$+ |\frac{1}{n} \sum_{k=0}^{n-1} m(E_0)m(F) - m(E)m(F)|$$

$$\leq 2\varepsilon + |\frac{1}{n} \sum_{k=0}^{n-1} m(T^{-k}E_0 \cap F_0) - m(E_0)m(F_0)| + \varepsilon + \varepsilon$$

for each n. (i) follows by the known behavior of the right hand side of this inequality.

To prove (ii) first show

$$|m(T^{-k}E \cap F) - m(E)m(F)| \leq 4\varepsilon + |m(T^{-k}E_0 \cap F_0) - m(E_0)m(F_0)| \qquad (*)$$

by leaving out " $\frac{1}{n} \sum_{k=0}^{n-1}$ " from the above inequalities, and then take Cesaro averages of each side of (*).

(iii) follows immediately from the inequality (*). //

Theorem 1.8:

If $\{a_n\}$ is a bounded sequence of real numbers then the following are equivalent:

(1)
$$\frac{1}{n} \sum_{i=0}^{n-1} |a_i| \to 0.$$

(2) $\exists \ J \subset Z^+$, J of density zero, i.e.,

$$\left(\frac{\text{cardinality } (J \cap \{0,1,\ldots,n-1\})}{n} \right) \to 0,$$

such that $\lim_n a_n = 0$ provided $n \notin J$.

(3)
$$\frac{1}{n} \sum_{i=0}^{n-1} |a_i|^2 \to 0.$$

Proof: If $M \subset Z^+$ let $\alpha_M(n)$ denote the cardinality of $\{0,1,\ldots,n-1\} \cap M$.

(1) $\Rightarrow$ (2). Let $J_k = \{n \in Z^+ : |a_n| \geq \frac{1}{k}\}$ (k > 0). Then $J_1 \subset J_2 \subset \ldots$. Each J_k has density zero since $\frac{1}{n} \sum_{i=0}^{n-1} |a_i| \geq \frac{1}{n} \frac{1}{k} \alpha_{J_k}(n)$. Therefore there exist integers $0 = \ell_0 < \ell_1 < \ell_2 < \ldots$ such that for $n \geq \ell_k$,

$$\frac{1}{n} \alpha_{J_{k+1}}(n) < \frac{1}{k+1} .$$

Set $J = \bigcup_{k=0}^{\infty} [J_{k+1} \cap [\ell_k, \ell_{k+1})]$. We now show that J has density zero. Since $J_1 \subset J_2 \subset \ldots,$ if $\ell_k \leq n < \ell_{k+1}$ we have

$$J \cap [0,n) = [J \cap [0,\ell_k)] \cup [J \cap [\ell_k, n)] \subset [J_k \cap [0,\ell_k)] \cup [J_{k+1} \cap [0,n)],$$

and therefore

$$\frac{1}{n} \alpha_J(n) \leq \frac{1}{n}[\alpha_{J_k}(\ell_k) + \alpha_{J_{k+1}}(n)] \leq \frac{1}{n}[\alpha_{J_k}(n) + \alpha_{J_{k+1}}(n)] < \frac{1}{k} + \frac{1}{k+1} .$$

Hence $\frac{1}{n} \alpha_J(n) \to 0$ as $n \to \infty,$ i.e., J has density zero. If
$n > \ell_k$ and $n \notin J$ then $n \notin J_{k+1}$ and therefore $|a_n| < \frac{1}{k+1}.$ Hence
$\lim_{J \not\ni n \to \infty} |a_n| = 0.$

(2) $\Rightarrow$ (1). Suppose $|a_n| \leq K \; \forall \, n.$ Let $\varepsilon > 0.$ There exists
N_ε such that $n \geq N_\varepsilon,$ $n \notin J$ imply $|a_n| < \varepsilon,$ and M_ε such that
$n \geq M_\varepsilon$ implies $\dfrac{\alpha_J(n)}{n} < \varepsilon.$ Then $n \geq \max(N_\varepsilon, M_\varepsilon)$ implies

$$\frac{1}{n} \sum_{i=0}^{n-1} |a_i| = \frac{1}{n}\left[\sum_{J \cap \{0,1,\ldots,n-1\}} |a_i| + \sum_{i \notin J \cap \{0,1,\ldots,n-1\}} |a_i| \right]$$

$$< \frac{K}{n} \alpha_J(n) + \varepsilon \; < \; (K+1)\varepsilon.$$

(1) $\leftrightarrow$ (3). By the above it suffices to note that $\lim_{J \not\ni n \to \infty} |a_n| = 0$
iff $\lim_{J \not\ni n \to \infty} |a_n|^2 = 0.$ //

Corollary 1.8:

T is weak-mixing iff $\forall \, A,B \in \mathcal{B} \; \exists \; J(A,B)$ of density zero in
$\mathbb{Z}^+ \; \ni \; \lim_{\substack{n \to \infty \\ n \notin J(A,B)}} m(T^{-n}A \cap B) = m(A)m(B)$ iff $\forall \, A,B \in \mathcal{B}$
$\frac{1}{n} \sum_{i=0}^{n-1} |m(T^{-i}A \cap B) - m(A)m(B)|^2 \to 0.$

Remark:

To say T is strong-mixing means that any set $B \in \mathcal{B}$ as it moves under T becomes, asymptotically, independent of a fixed set A ∈ B. T is weak mixing means B becomes independent of A if we neglect a few instants of time. T is ergodic means B becomes independent of A on the average.

The next result expresses the mixing concepts in functional form. Recall that U_T is defined on functions by $U_T f = f \circ T$.

Theorem 1.9:

Suppose (X, B, m) is a probability space and $T: X \to X$ is measure-preserving. Then

(a) T is ergodic iff for all $f, g \in L^2(m)$

$$\frac{1}{n} \sum_{i=0}^{n-1} (U_T^i f, g) \to (f, 1)(1, g)$$

iff for all $f \in L^2(m)$

$$\frac{1}{n} \sum_{i=0}^{n-1} (U_T^i f, f) \to (f, 1)(1, f).$$

(b) T is weak-mixing iff for all $f, g \in L^2(m)$

$$\frac{1}{n} \sum_{i=0}^{n-1} |(U_T^i f, g) - (f, 1)(1, g)| \to 0$$

iff for all $f \in L^2(m)$

$$\frac{1}{n} \sum_{i=0}^{n-1} |(U_T^i f, f) - (f, 1)(1, f)| \to 0$$

iff for all $f \in L^2(m)$

$$\frac{1}{n} \sum_{i=0}^{n-1} |(U_T^i f, f) - (f, 1)(1, f)|^2 \to 0.$$

(c) (1) T is strong-mixing iff

(2) for all $f, g \in L^2(m)$, $(U_T^n f, g) \to (f, 1)(1, g)$

iff (3) for all $f \in L^2(m)$, $(U_T^n f, f) \to (f, 1)(1, f).$

Proof: (a), (b), and (c) are proved using similar methods. We shall prove (c) to illustrate the ideas. Slight modification of this proof will prove (a) and (b).

(2) $\Rightarrow$ (1). This follows by putting $f = \chi_A$, $g = \chi_B$, for $A, B \in \mathcal{B}$.

(1) $\Rightarrow$ (3). We easily get that for any $A, B \in \mathcal{B}$, $(U_T^n \chi_A, \chi_B) \rightarrow (\chi_A, 1)(1, \chi_B)$. Fixing B, we get that $(U_T^n h, \chi_B) \rightarrow (h, 1)(1, \chi_B)$ for any simple function h. Then, fixing h, we get that $(U_T^n h, h) \rightarrow (h, 1)(1, h)$. So (3) is true for all simple functions.

Suppose $f \in L^2(m)$, and let $\varepsilon > 0$. Choose a simple function h $\ni$ $\|f - h\|_2 < \varepsilon$, and choose $N(\varepsilon)$ so that $n \geq N(\varepsilon)$ implies $|(U_T^n h, h) - (h, 1)(1, h)| < \varepsilon$. Then if $n \geq N(\varepsilon)$

$$|(U_T^n f, f) - (f, 1)(1, f)| \leq |(U_T^n f, f) - (U_T^n h, f)|$$

$$+ |(U_T^n h, f) - (U_T^n h, h)| + |(U_T^n h, h) - (h, 1)(1, h)|$$

$$+ |(h, 1)(1, h) - (f, 1)(1, h)| + |(f, 1)(1, h) - (f, 1)(1, f)|$$

$$\leq |(U_T(f-h), f)| + |(U_T h, f-h)|$$

$$+ \varepsilon + |(1, h)||(h-f, 1)| + |(f, 1)||(1, h-f)|$$

$$\leq \|f - h\|_2 \|f\|_2 + \|f-h\|_2 \|h\|_2 + \varepsilon + \|h\|_2 \|f-h\|_2 + \|f\|_2 \|h-f\|_2$$

$$\text{by the Schwartz inequality}$$

$$\leq \varepsilon \|f\|_2 + \varepsilon(\|f\|_2 + \varepsilon) + \varepsilon + (\|f\|_2 + \varepsilon)\varepsilon + \varepsilon \|f\|_2 .$$

Therefore $(U_T^n f, f) \rightarrow (f, 1)(1, f)$ as $n \rightarrow \infty$.

(3) $\Rightarrow$ (2). Let $f \in L^2(m)$ and let H_f denote the smallest (closed) subspace of $L^2(m)$ containing f and the constant functions and satisfying $U_T H_f \subset H_f$.

$$F_f = \{g \in L^2(m) : (U_T^n f, g) \rightarrow (f, 1)(1, g)\}$$

is a subspace of $L^2(m)$, is closed, contains f and the constant

44

functions and is U_T invariant so it contains H_f. If $g \perp H_f$ then
$(U_T^n f, g) = 0$ for $n \geq 0$ and $(1,g) = 0$ and therefore $H_f^\perp \subset F_f$.
Hence $F_f = L^2(m)$. //

Definition 1.5:

If $T: X \to X$ is measure-preserving, define $T \times T: X \times X \to X \times X$ by
$(T \times T)(x,y) = (T(x), T(y))$. This is a measure-preserving transformation
on $(X \times X, B \times B, m \times m)$ by Theorem 1.1 since it is measure-preserving on
measurable rectangles and hence on finite disjoint unions of such
rectangles.

Theorem 1.10:

If T is a measure-preserving transformation on a probability
space X then the following are equivalent:
(1) T is weak-mixing.
(2) $T \times T$ is ergodic.
(3) $T \times T$ is weak-mixing.

Proof: (1) $\Rightarrow$ (3). Let $A, B \in B$, $C, D \in B$. $\exists$ J_1, J_2 of density
zero such that

$$\lim_{\substack{n \notin J_1 \\ n \to \infty}} m(T^{-n}A \cap B) = m(A)m(B)$$

$$\lim_{\substack{n \notin J_2 \\ n \to \infty}} m(T^{-n}C \cap D) = m(C)m(D).$$

Then

$$\lim_{\substack{n \notin J_1 \cup J_2 \\ n \to \infty}} (m \times m)\{(T \times T)^{-n}(A \times C) \cap (B \times D)\} = \lim_{\substack{n \notin J_1 \cup J_2 \\ n \to \infty}} m(T^{-n}A \cap B)m(T^{-n}C \cap D)$$

$$= m(A)m(B)m(C)m(D)$$

$$= (m \times m)(A \times C)(m \times m)(C \times D).$$

Thus the proper relationship holds for rectangles and hence for finite
disjoint unions of these rectangles. These we know form an algebra F

which generates the σ-algebra $\mathcal{B}$. By Corollary 1.8 we have

$$\frac{1}{n} \sum_{i=0}^{n-1} |m(T^{-i}A \cap B) - m(A)m(B)| \to 0 \quad \forall \ A,B \in F \quad \text{and the result follows}$$

by Theorem 1.7.

(3) $\Rightarrow$ (2) is clear.

(2) $\Rightarrow$ (1). Let $A,B \in \mathcal{B}$. We have that

$$\frac{1}{n} \sum_{i=0}^{n-1} m(T^{-i}A \cap B) = \frac{1}{n} \sum_{i=0}^{n-1} (m \times m)((T \times T)^{-i}(A \times X) \cap (B \times X))$$

$$\to (m \times m)(A \times X)(m \times m)(B \times X) \quad \text{by (2)}$$

$$= m(A)m(B).$$

Also

$$\frac{1}{n} \sum_{i=0}^{n-1} (m(T^{-i}A \cap B))^2 = \frac{1}{n} \sum_{i=0}^{n-1} (m \times m)((T \times T)^{-i}(A \times A) \cap (B \times B))$$

$$\to (m \times m)(A \times A)(m \times m)(B \times B) \quad \text{by (2)}$$

$$= m(A)^2 m(B)^2.$$

Thus

$$\frac{1}{n} \sum_{i=0}^{n-1} \{m(T^{-i}A \cap B) - m(A)m(B)\}^2$$

$$= \frac{1}{n} \sum_{i=0}^{n-1} \{m(T^{-i}A \cap B)^2 - 2m(T^{-i}A \cap B)m(A)m(B) + m(A)^2 m(B)^2\}$$

$$\to 2m(A)^2 m(B)^2 - 2m(A)^2 m(B)^2 = 0.$$

Therefore T is weak-mixing by Corollary 1.8. //

Definition 1.6:

Let $T: (X,\mathcal{B},m) \to (X,\mathcal{B},m)$ be a measure-preserving transformation on a probability space. We say that λ is an eigenvalue of T, ($\lambda \in C$) if $\exists \ f \neq 0 \in L^2(m) \ni U_T f = \lambda f$ in $L^2(m)$; i.e., $f(Tx) = \lambda f(x)$ a.e. We call f an eigenfunction corresponding to λ.

46

Remarks:

(i) If λ is an eigenvalue of T then $|\lambda| = 1$ since

$$\|f\|^2 = \|U_T f\|^2 = (U_T f, U_T f) = (\lambda f, \lambda f) = |\lambda|^2 \|f\|^2.$$

(ii) $\lambda = 1$ is always an eigenvalue corresponding to any constant eigenfunction.

Definition 1.7:

   We say that $T: X \to X$ has continuous spectrum if 1 is the only eigenvalue of T and the only eigenfunctions are the constants.

   Observe that T has continuous spectrum iff $\lambda = 1$ is the only eigenvalue and T is ergodic.

   We shall need the following result from spectral theory to prove the next theorem. The proof can be found in Halmos [3].

Spectral Theorem for Unitary Operators:

   Suppose U is a unitary operator on a complex Hilbert space H. Then $\forall f \in H$, $\exists$ a unique finite Borel measure μ_f on K $\ni$

$$(U^n f, f) = \int_K \lambda^n d\mu_f(\lambda) \quad \forall n \in Z.$$

   If T is an invertible measure-preserving transformation then U_T is unitary, and if T has continuous spectrum then μ_f has no atoms for all $f \in L^2(m)$ with $(f,1) = 0$.

Theorem 1.11:

   If T is an invertible measure-preserving transformation of a probability space then T is weak-mixing iff T has continuous spectrum.

   Proof: ($\Rightarrow$). Suppose $fT = \lambda f$ a.e., $f \in L^2(m)$. If $\lambda \neq 1$ then integration gives $(f,1) = 0$ and by the weak-mixing property

$$\frac{1}{n} \sum_{i=0}^{n-1} |(U_T^i f, f)| \rightarrow 0$$

i.e.,

$$\frac{1}{n} \sum_{i=0}^{n-1} |(\lambda^i f, f)| \rightarrow 0.$$

Since $|\lambda^i| = 1$, this gives $(f,f) = 0$ and therefore $f = 0$ a.e.
If $\lambda = 1$ then f = constant a.e. by the ergodicity of T.

(⇐). Suppose T has continuous spectrum. We show that if
$f \in L^2(m)$ then

$$\frac{1}{n} \sum_{i=0}^{n-1} |(U_T^i f, f) - (f,1)(1,f)|^2 \rightarrow 0.$$

If f is constant a.e. this is true. Hence, all we need to show is
that $(f,1) = 0$ implies

$$\frac{1}{n} \sum_{i=0}^{n-1} |(U_T^i f, f)|^2 \rightarrow 0.$$

By the spectral theorem it suffices to show that if μ_f is a continuous (non-atomic) measure on K then

$$\frac{1}{n} \sum_{i=0}^{n-1} |\int \lambda^i d\mu_f(\lambda)|^2 \rightarrow 0.$$

We have

$$\frac{1}{n} \sum_{i=0}^{n-1} |\int \lambda^i d\mu_f(\lambda)|^2 = \frac{1}{n} \sum_{i=0}^{n-1} \left(\int \lambda^i d\mu_f(\lambda) \cdot \int \lambda^{-i} d\mu_f(\lambda) \right)$$

$$= \frac{1}{n} \sum_{i=0}^{n-1} \left(\int \lambda^i d\mu_f(\lambda) \cdot \int \tau^{-i} d\mu_f(\tau) \right)$$

$$= \frac{1}{n} \sum_{i=0}^{n-1} \iint_{K \times K} (\lambda\bar{\tau})^i d(\mu_f \times \mu_f)(\lambda,\tau) \quad \text{(by Fubini's Theorem)}$$

$$= \iint_{K \times K} \left(\frac{1}{n} \sum_{i=0}^{n-1} (\lambda\bar{\tau})^i \right) d(\mu_f \times \mu_f)(\lambda,\tau).$$

If (λ, τ) is not in the diagonal of $K \times K$ then

$$\frac{1}{n} \sum_{i=0}^{n-1} (\lambda \bar{\tau})^i = \frac{1}{n} \left[\frac{1 - (\lambda \bar{\tau})^n}{1 - (\lambda \bar{\tau})} \right] \to 0$$

as $n \to \infty$. Since μ_f has no atoms $(\mu_f \times \mu_f)(\text{diagonal}) = 0$ and there-fore the integrand $\to 0$ a.e. The integrand has modulus ≤ 1, so that we can apply the bounded convergence theorem to obtain the result. //

We now investigate the mixing properties of the examples men-tioned in §1.

Examples:

(1) I = identity on $(X, \mathcal{B}, m)$. I is ergodic iff all the elements of $\mathcal{B}$ have measure 0 or 1 iff I is strong-mixing.

(2) $T(z) = az$ on K. T is never weak-mixing since if $f(z) = z$ then $f(Tz) = f(az) = af(z)$ and $f \neq$ constant. (This has used the trivial part of Theorem 1.11.)

(3) No rotation on a compact group is weak-mixing. We have already mentioned that if T is ergodic then the group G is abelian; and then if $Tx = ax$ and γ is any character of G we have $\gamma(Tx) = \gamma(a)\gamma(x)$, which shows that T does not have continuous spectrum.

(4) Endomorphisms of compact metric groups are strong-mixing iff weak-mixing iff ergodic.

Proof: We shall give the proof when G is abelian. It suf-fices to show that if the endomorphism $A: G \to G$ is ergodic then A is strong-mixing. If $\gamma, \delta \in \hat{G}$ then $(U_A^n \gamma, \delta) = 0$ eventually unless $\gamma = \delta \equiv 1$. So always $(U_A^n \gamma, \delta) \to (\gamma, 1)(1, \delta)$. Fix $\delta \in \hat{G}$. The col-lection

$$H_\delta = \{f \in L^2(m): (U_A^n f, \delta) \to (f, 1)(1, \delta)\}$$

is a closed subspace of $L^2(m)$. (To check H_δ is closed, suppose

$f_k \in H$ and $f_k \to f \in L^2(m)$. For $\delta \equiv 1$ it is clear that $H_\delta = L^2(m)$. So suppose $(1,\delta) = 0$. Then

$$|(U_A^n f,\delta)| \le |(U_A^n f,\delta) - (U_A^n f_k,\delta)| + |(U_A^n f_k,\delta)|$$

$$\le \|f-f_k\|_2 \|\delta\|_2 + |(U_A^n f_k,\delta)| \quad \text{(by the Schwarz inequality)}$$

$$= \|f-f_k\|_2 + |(U_A^n f_k,\delta)| \quad < \quad \varepsilon$$

if $n \ge N(\varepsilon)$ where k is chosen so that $\|f-f_k\|_2 < \varepsilon/2$ and $N(\varepsilon)$ is chosen so that $n \ge N(\varepsilon)$ implies $|(U_A^n f_k,\delta)| < \varepsilon/2$.) Since H_δ contains $\hat{G}$, it is equal to $L^2(m)$. Fix $f \in L^2(m)$ and consider $L_f = \{g \in L^2(m): (U_A^n f,g) \to (f,1)(1,g)\}$. L_f is a closed subspace of $L^2(m)$, contains $\hat{G}$ by the above, and so equals $L^2(m)$. Hence A is strong-mixing. //

(5) An affine transformation $T = a \cdot A$ on a compact metric abelian group is strong-mixing iff it is weak-mixing iff A is ergodic.

   <u>Proof</u>: We shall give the proof in the case when G is connected. Let $Bx = x^{-1}A(x)$ and recall that T is ergodic iff

   (i) $\gamma \circ A^k = \gamma$ $k > 0$ implies $\gamma \circ A = \gamma$, and

   (ii) $[a, BG] = G$.

   If A is ergodic then $BG = G$ since the endomorphism $\hat{B}$ of $\hat{G}$ is one-to-one. Choose $b \in G$ so that $B(b) = a$. Put $\phi(x) = bx$; then $\phi T = A\phi$. ϕ preserves Haar measure m on G and hence induces a unitary operator U_ϕ on $L^2(m)$. We then have that U_T and U_A are isomorphic as Hilbert space operators by the induced conjugacy

$$U_T \circ U_\phi = U_\phi \circ U_A.$$

Now A is strong-mixing by (4), and hence U_A satisfies the conditions of part (c) of Theorem 1.9. But then U_T satisfies these conditions, and hence T is strong-mixing.

Conversely if T is strong-mixing and A is not ergodic then by (i) $\gamma \circ A = \gamma$ for some $\gamma \not\equiv 1$. But then

$$|(U_T^n \gamma, \gamma)| = |(\gamma(a)\gamma(Aa)\ldots\gamma(A^{n-1}a)\gamma, \gamma)| = \|\gamma\|_2^2 = 1$$

which does not converge to $|(\gamma,1)(1,\gamma)| = 0$, contradicting the mixing of T. So if T is strong-mixing then A is ergodic. //

(6) The 2-sided $(p_0,\ldots,p_{k-1})$-shift is strong-mixing. This is proven by doing the easy verification on measurable rectangles, then on their disjoint finite unions, and then applying Theorem 1.7.

(7) Similarly, the 1-sided $(p_0,\ldots,p_{k-1})$-shift is strong-mixing.

Chapter 2: Isomorphism and Spectral Invariants

§1. Isomorphism of Measure-Preserving Transformations

What should we mean by saying that two measure-preserving transformations are the "same"? We must bear in mind that sets of measure
0 do not matter from the point of view of measure theory.

Examples:

(1) Let T be the transformation $Tz = z^2$ on the unit circle K
with Borel sets and Haar measure, and let S be given by $Sx = 2x$
mod 1 on [0,1) with Borel sets and Lebesgue measure. Consider the
map ϕ: [0,1) → K defined by $x \to e^{2\pi i x}$. ϕ is a bijection and preserves measure (check on finite unions of intervals and use Theorem
1.1). Also $\phi S = T\phi$. So, we want to regard T and S as the "same".

(2) Again, let S be the transformation $Sx = 2x$ mod 1 on [0,1)
with Borel sets and Lebesgue measure, and let T_2: X → X be the
1-sided (½,½)-shift. Define ψ: X → [0,1) by

$$\psi(a_1, a_2, a_3, \ldots) = \frac{a_1}{2} + \frac{a_2}{2^2} + \frac{a_3}{2^3} + \ldots \quad .$$

ψ is not one-to-one only at points $(a_1, a_2, \ldots)$ whose coordinates are
constant eventually. ψ, though, is onto and $\psi T_2 = S\psi$. Also ψ preserves measure; we can check this out on dyadic intervals and then on
their finite disjoint unions and apply Theorem 1.1.

Suppose D is the set of points of the space X of the 1-sided
(½,½)-shift which have constant coordinates eventually. Then $T_2^{-1}D = D$
and so $T_2^{-1}(X \backslash D) = X \backslash D$. Let D_2 consist of the dyadic rationals in
[0,1). Then $S^{-1}D_2 = D_2$, so that $S^{-1}([0,1) \backslash D_2) = [0,1) \backslash D_2$.

We see that the diagram

$$\begin{array}{ccc} X \backslash D & \xrightarrow{\ \ T_2\ \ } & X \backslash D \\ \Big\downarrow & & \Big\downarrow \quad \text{1:1 onto} \\ [0,1)\backslash D_2 & \xrightarrow{\ \ S\ \ } & [0,1)\backslash D_2 \end{array}$$

commutes.

We would like to consider these transformations as isomorphic since, after removing sets of measure zero, we can throw one to the other by an invertible measure-preserving transformation.

Definition 2.1:

Suppose (X_1, B_1, m_1) and (X_2, B_2, m_2) are probability spaces together with measure-preserving transformations $T_1: X_1 \to X_1$, $T_2: X_2 \to X_2$. We say that T_1 is <u>isomorphic</u> to T_2 if $\exists\ M_1 \in B_1$, $m_1(M_1) = 1$, $M_2 \in B_2$, $m_2(M_2) = 1$ $\ni$

  (i) $T_1 M_1 \subseteq M_1$, $T_2 M_2 \subseteq M_2$, and

  (ii) $\exists$ an invertible measure-preserving transformation
      $\phi: M_1 \to M_2 \ni \phi T_1(x) = T_2 \phi(x)\ \forall\ x \in M_1$.

We write $T_1 \simeq T_2$. (In (ii) the set M_i $(i = 1,2)$ is assumed to be equipped with the σ-algebra $M_i \cap B_i = \{M_i \cap B \mid B \in B_i\}$ and the restriction of the measure m_i to this σ-algebra.)

Remarks:

(a) $\simeq$ is an equivalence relation.

(b) $T_1 \simeq T_2 \Rightarrow T_1^n \simeq T_2^n$, $\forall\ n > 0$.

(c) If T_1 and T_2 are invertible we can take M_1, M_2 so that $T_1 M_1 = M_1$, $T_2 M_2 = M_2$; we just take $\bigcap_{-\infty}^{\infty} T_1^k M_1$, $\bigcap_{-\infty}^{\infty} T_2^k M_2$ as the new sets.

§2. Conjugacy of Measure-Preserving Transformations

Although the notion of isomorphism, introduced above, is useful in practice the following is mathematically more natural.

Given (X, B, m) we define an equivalence relation on B by saying that $A \sim B$ iff $m(A \Delta B) = 0$. Let $\underset{\sim}{B}$ denote the set of equivalence classes. $\underset{\sim}{B}$ is a Boolean σ-algebra under the operations induced from the usual operations on B. m induces a measure $\underset{\sim}{m}$ on $\underset{\sim}{B}$. We call $(\underset{\sim}{B}, \underset{\sim}{m})$ a __measure algebra__. Note that for $\underset{\sim}{B} \in \underset{\sim}{B}$, $\chi_{\underset{\sim}{B}}$ is a uniquely defined member of $L^2(m)$.

Suppose $T: X \to X$ is measure-preserving. If $A \sim B$ then $T^{-1}A \sim T^{-1}B$; so we have a map $\underset{\sim}{T}^{-1}: \underset{\sim}{B} \to \underset{\sim}{B}$ which is defined by $\underset{\sim}{T}^{-1}(\underset{\sim}{B}) = \underset{\sim}{T^{-1}B}$. $\underset{\sim}{T}^{-1}$ preserves unions, intersections, and complements, and $\underset{\sim}{m}(\underset{\sim}{T}^{-1}\underset{\sim}{B}) = \underset{\sim}{m}(\underset{\sim}{B})$.

Definition 2.2:

A map $\Phi: (\underset{\sim}{B}_2, \underset{\sim}{m}_2) \to (\underset{\sim}{B}_1, \underset{\sim}{m}_1)$ of measure algebras is called an __isomorphism of measure algebras__ if Φ is a surjective bijection and preserves complements and countable unions and

$$\underset{\sim}{m}_1(\Phi(\underset{\sim}{B}_2)) = \underset{\sim}{m}_2(\underset{\sim}{B}_2) \quad \forall \, \underset{\sim}{B}_2 \in \underset{\sim}{B}_2.$$

Definition 2.3:

We say that $T_1: X_1 \to X_1$, $T_2: X_2 \to X_2$ are __conjugate__ if $\exists$ a measure algebra isomorphism $\Phi: (\underset{\sim}{B}_2, \underset{\sim}{m}_2) \to (\underset{\sim}{B}_1, \underset{\sim}{m}_1)$ such that $\Phi \underset{\sim}{T}_2^{-1} = \underset{\sim}{T}_1^{-1} \Phi$.

Remarks:

(1) Conjugacy is an equivalence relation.

(2) $T_1 \simeq T_2 \Rightarrow T_1$ and T_2 are conjugate.

Just let $\Phi = \underset{\sim}{\phi}^{-1}$, which is uniquely defined although ϕ is not defined on the whole of X_1.

In Lebesgue spaces (i.e., probability spaces isomorphic to a subinterval of [0,1] with Lebesgue measure possibly together with
countably many atoms) conjugacy implies isomorphism. In particular,
a compact separable metric space with a completed Borel measure is a
Lebesgue space.

§3. Spectral Isomorphism

Suppose T: X → X is a measure-preserving transformation on a
probability space (X,B,m). We have defined U_T: $L^2(m)$ → $L^2(m)$ by
f ↦ f∘T, and noted that ∀ f,g ∈ $L^2(m)$ we have $(U_T f, U_T g) = (f,g)$.
Also, if T is one-to-one, U_T is unitary. A spectral property
of T is a property of U_T.

Definition 2.4:

Measure-preserving transformations T_1 on (X_1,B_1,m_1), and T_2
on (X_2,B_2,m_2) are spectrally isomorphic if ∃ a linear operator
W: $L^2(m_2)$ → $L^2(m_1)$ such that

(i) W is invertible

(ii) (Wf,Wg) = (f,g) ∀ f,g ∈ $L^2(m_2)$

(iii) $U_{T_1} W = W U_{T_2}$.

(The conditions (i), (ii) just say that W is an isomorphism of Hilbert spaces.)

Remarks:

(1) Spectral isomorphism is an equivalence relation.

(2) If Φ: (B_2,m_2) → (B_1,m_1) is a measure algebra isomorphism then
Φ induces an invertible linear map V: $L^2(m_2)$ → $L^2(m_1)$, by
$V\chi_B = \chi_{\Phi(B)}$, with the properties:

(a) (Vf,Vg) = (f,g) ∀ f,g ∈ $L^2(m_2)$.

(b) V, V^{-1} map bounded functions to bounded functions.

(c) V is multiplicative on bounded functions.

Proof: V is defined as follows. Let $B_2 \in B_2$; then $V(x_{B_2}) = x_{\Phi(B_2)}$ which is unambiguous in $L^2(m_2)$. We then extend V to simple functions and then to $L^2(m_2)$ functions. The properties of Φ guarantee this can be done. (a), (b), and (c) are proved by checking first for characteristic functions, then for simple functions, and then extending to the whole of $L^2(m_2)$. //

(3) If T_1 and T_2 are conjugate then they are spectrally isomorphic.

Proof: Suppose $\Phi: (B_2, m_2) \rightarrow (B_1, m_1)$ is an isomorphism of measure algebras such that $\Phi T_2^{-1} = T_1^{-1} \Phi$. Let V be defined as in remark (2). It remains to check that

$$VU_{T_2} = U_{T_1} V.$$

First, on characteristic functions

$$U_{T_1} V(x_{B_2}) = U_{T_1}(x_{\Phi B_2}) = x_{T_1^{-1}\Phi B_2} = x_{\Phi T_2^{-1} B_2} = V(x_{T_2^{-1}B_2}) = VU_{T_2}(x_{B_2}).$$

Therefore $U_{T_1} V$ and VU_{T_2} agree on characteristic functions and hence on linear combinations of characteristic functions. By their continuity we have $U_{T_1} V = VU_{T_2}$. //

The following tells us when spectral isomorphism implies conjugacy.

Theorem 2.1:

An invertible isometry $V: L^2(m_2) \rightarrow L^2(m_1)$ is induced by an isomorphism of measure algebras (in the same sense of remark (2)) if both V and V^{-1} take bounded functions to bounded functions and $V(fg) = V(f)V(g)$ whenever f and g are bounded and in $L^2(m_2)$.

Proof: Let $B_2 \in \mathcal{B}_2$. We have $x_{B_2}^2 = x_{B_2}$ so that

$$V(x_{B_2}^2) = V(x_{B_2})V(x_{B_2}) = V(x_{B_2}),$$

and we see that $V(x_{B_2})$ takes 1 and 0 as its only values. Thus $\exists\ B_1 \in \mathcal{B}_1$ such that $V(x_{B_2}) = x_{B_1}$ a.e. We define $\Phi \colon (\underset{\sim}{\mathcal{B}}_2, \underset{\sim}{m}_2) \to (\underset{\sim}{\mathcal{B}}_1, \underset{\sim}{m}_1)$ by $\Phi(\underset{\sim}{B}_2) = \underset{\sim}{B}_1$. This is unambiguous since if $m_2(B_2 \triangle A_2) = 0$ then $\|x_{B_2} - x_{A_2}\| = 0$ so that $\|V(x_{B_2}) - V(x_{A_2})\| = 0$. Clearly, V is induced by Φ in the sense of remark 2.

We now show that Φ is an isomorphism of measure algebras. First, Φ is invertible by doing the above for V^{-1}. Also,

$$\underset{\sim}{m}_2(\underset{\sim}{B}_2) = m_2(B_2) = \int x_{B_2} \bar{x}_{B_2}\ dm_2 = (x_{\underset{\sim}{B}_2}, x_{\underset{\sim}{B}_2})$$

$$= (Vx_{\underset{\sim}{B}_2}, Vx_{\underset{\sim}{B}_2}) = (x_{\Phi(\underset{\sim}{B}_2)}, x_{\Phi(\underset{\sim}{B}_2)}) = \underset{\sim}{m}_1(\Phi\underset{\sim}{B}_2).$$

It remains to show that Φ preserves complements and countable unions. First note that since V is norm-preserving and maps characteristic functions to characteristic functions, $V(1) = 1$.

Since $x_{\underset{\sim}{B}_2} + x_{\underset{\sim}{X}_2 \setminus \underset{\sim}{B}_2} = 1$ in $L^2(m_2)$ applying V to both sides gives $x_{\Phi\underset{\sim}{B}_2} + x_{\Phi(\underset{\sim}{X}_2 \setminus \underset{\sim}{B}_2)} = 1$ so $\underset{\sim}{X}_1 \setminus \Phi\underset{\sim}{B}_2 = \Phi(\underset{\sim}{X}_2 \setminus \underset{\sim}{B}_2)$. Therefore Φ preserves complements.

Suppose $\underset{\sim}{B}, \underset{\sim}{C} \in \underset{\sim}{\mathcal{B}}_2$. Then

$$x_{\underset{\sim}{B} \cup \underset{\sim}{C}} = x_{\underset{\sim}{B}} + x_{\underset{\sim}{C}} - x_{\underset{\sim}{B} \cap \underset{\sim}{C}} = x_{\underset{\sim}{B}} + x_{\underset{\sim}{C}} - x_{\underset{\sim}{B}} x_{\underset{\sim}{C}}.$$

Taking V of both sides we get:

$$x_{\Phi(\underset{\sim}{B} \cup \underset{\sim}{C})} = x_{\Phi(\underset{\sim}{B})} + x_{\Phi(\underset{\sim}{C})} - x_{\Phi(\underset{\sim}{B})} x_{\Phi(\underset{\sim}{C})} = x_{\Phi(\underset{\sim}{B}) \cup \Phi(\underset{\sim}{C})}.$$

Thus $\Phi(\underset{\sim}{B} \cup \underset{\sim}{C}) = \Phi(\underset{\sim}{B}) \cup \Phi(\underset{\sim}{C})$ and hence, (by induction) preserves all finite unions.

Let $B_1, B_2, \ldots, B_n, \ldots \in B_2$, then

$$X_{\bigcup_{i=1}^{n} B_i} \rightarrow X_{\bigcup_{i=1}^{\infty} B_i} \qquad \text{a.e.}$$

and also in $L^2(m_2)$ by the bounded convergence theorem. Since V is an isometry it is continuous, so,

$$V\left(X_{\bigcup_{i=1}^{n} B_i}\right) \rightarrow V\left(X_{\bigcup_{i=1}^{\infty} B_i}\right) = X_{\Phi\left(\bigcup_{i=1}^{\infty} B_i\right)} \qquad \text{in } L^2(m_1).$$

On the other hand,

$$V\left(X_{\bigcup_{i=1}^{n} B_i}\right) = X_{\Phi\left(\bigcup_{i=1}^{n} B_i\right)} = X_{\bigcup_{i=1}^{n} \Phi B_i}$$

by the above and so converges to $X_{\bigcup_{i=1}^{\infty} \Phi B_i}$ in $L^2(m_1)$. Therefore

$$\Phi(\bigcup_{i=1}^{\infty} B_i) = \bigcup_{i=1}^{\infty} \Phi B_i. \quad //$$

Corollary 2.1:

If $T_1: X_1 \rightarrow X_1$, $T_2: X_2 \rightarrow X_2$ are measure-preserving and if $U_{T_1} V = V U_{T_2}$ for $V: L^2(m_2) \rightarrow L^2(m_1)$ satisfying the conditions of Theorem 2.1, then T_1 and T_2 are conjugate.

§4. Spectral Invariants

Definition 2.5:

A property P of measure-preserving transformations is a

$\begin{bmatrix} \text{isomorphism} \\ \text{conjugacy} \\ \text{spectral} \end{bmatrix}$ invariant if the following holds:

Given T_1 has P and T_2 is $\begin{bmatrix} \text{isomorphic} \\ \text{conjugate} \\ \text{spectrally isomorphic} \end{bmatrix}$ to T_1,

then T_2 has property P.

Note:

A spectral invariant is a conjugacy invariant, and a conjugacy invariant is an isomorphism invariant.

Theorem 2.2:

The following are spectral invariants of measure-preserving transformations:

(i) Ergodicity

(ii) Weak-mixing

(iii) Strong-mixing.

Proof: (i) T is ergodic iff $\{f \in L^2(m): U_T f = f\}$ is a one-dimensional subspace.

(ii) T is weak-mixing iff 1 is the only eigenvalue and T is ergodic.

(iii) Suppose $WU_{T_2} = U_{T_1} W$ and T_1 is strong-mixing. We have to show that

$$(U_{T_2} h, k) \to (h, 1)(1, k) \quad \forall h, k \in L^2(m_2).$$

This is true if h is constant or if k is constant, so assume $(h, 1) = 0 = (k, 1)$. Since T_1 is ergodic then T_2 is ergodic by (i) and since W sends the invariant functions for T_2 onto those for T_1, W maps the subspace of constants in $L^2(m_2)$ onto the subspace of constants in $L^2(m_1)$. So $(Wh, 1) = 0 = (1, Wk)$. Since W preserves the inner product,

$$(U_{T_2} h, k) = (WU_{T_2} h, Wk) = (U_{T_1} Wh, Wk) \to 0$$

since T_1 is strong-mixing. Therefore T_2 is strong-mixing. //

§5. Examples

Recall that T_1 is isomorphic to T_2

$\Rightarrow$ T_1 is conjugate to T_2

$\Rightarrow$ T_1 is spectrally isomorphic to T_2

and the converse of the first implication holds in all "decent"
measure spaces.

(1) Consider $T_1, T_2: K \to K$ given by $T_1(z) = a_1 z$, $T_2(z) = a_2 z$
where a_1 is a root of unity and a_2 is not a root of unity. T_i is
not ergodic while T_2 is ergodic. Hence they cannot be spectrally
isomorphic.

(2) Let $T(z) = az$ where a is not a root of unity. We know that
T is ergodic but not weak-mixing. Consider $A: T^2 \to T^2$ defined by
$A(z,w) = (zw,z)$. Since none of the eigenvalues of the matrix $\begin{pmatrix} 1 & 1 \\ 1 & 0 \end{pmatrix}$
are roots of unity, A is weak-mixing. Hence T and A are not
spectrally isomorphic.

(3) Let at least two of the numbers $\{p_1, p_2, \ldots, p_n\}$ be non-zero, and
$\sum_{i=1}^{n} p_i = 1$. Let the same be true for the numbers $\{q_1, \ldots, q_m\}$. We
claim that the 2-sided $(p_1, \ldots, p_n)$-shift and the 2-sided $(q_1, \ldots, q_m)$-
shift are spectrally isomorphic but not necessarily conjugate. A con-
sideration of entropy shows that they need not be conjugate. (See
Chapter 4.)

Consider the special case of the $(\frac{1}{2}, \frac{1}{2})$-shift T, with
$X = \prod_{-\infty}^{\infty} \{-1, 1\}$. A basis for $L^2(\{-1,1\})$ consists of the constant
function 1 and the identity map sending

$$(-1) \mapsto (-1), \quad 1 \mapsto 1.$$

Moreover, $L^2(\prod X_i)$ is the tensor product of the spaces $L^2(X_i)$ so
that we have an orthonormal basis for $L^2(X)$ consisting of all

functions $X \to C$ of the form:

$$g_0(\{x_n\}) = 1$$

and, for $n_1 < n_2 < \cdots < n_r$

$$g_{n_1,\ldots,n_r}(\{x_n\}) = x_{n_1} \cdot x_{n_2} \cdot \cdots \cdot x_{n_r} \; .$$

Note that

$$U_T g_{n_1,\ldots,n_r}(\{x_n\}) = (g_{n_1,\ldots,n_r} \circ T)(\{x_n\})$$

$$= x_{n_1+1} \cdot x_{n_2+1} \cdot \cdots \cdot x_{n_r+1} = g_{n_1+1,n_2+1,\ldots,n_r+1}(\{x_n\}),$$

that is,

$$U_T g_{n_1,\ldots,n_r} = g_{n_1+1,\ldots,n_r+1} .$$

So we have an orthonormal basis for $L^2(X)$ of the form;

$$f_0 \equiv 1, \quad \{U_T^n f_1\}_{n \in Z}, \quad \{U_T^n f_2\}_{n \in Z}, \quad \cdots .$$

Diagramatically, the basis has the form

$$f_0 \equiv 1$$

$$\cdots, \; U_T^{-2} f_1, \; U_T^{-1} f_1, \; f_1, \; U_T f_1, \; U_T^2 f_1, \; \cdots$$

$$\cdots, \; U_T^{-2} f_2, \; U_T^{-1} f_2, \; f_2, \; U_T f_2, \; U_T^2 f_2, \; \cdots \qquad (*)$$

$$\vdots \qquad \vdots \qquad \vdots \qquad \vdots \qquad \vdots$$

Definition 2.6:

An invertible measure-preserving transformation $T: X \to X$ has
countable Lebesgue spectrum if there exists an orthonormal basis for
$L^2(X)$ of the form:

$$f_0 \equiv 1, \quad \{U_T^n f_j\} \quad j \geq 1, \quad n \in Z,$$

i.e., a basis as in $(*)$ above.

Remarks:

(1) Any two transformations with countable Lebesgue spectrum are spectrally isomorphic.

Proof: If $T: X \to X$, $S: Y \to Y$ have bases:

$$f_0 \equiv 1, \quad \{U_T^n f_j\}_{j \in Z^+}^{n \in Z} \quad \text{for} \quad L^2(X)$$

$$g_0 \equiv 1, \quad \{U_S^n g_j\}_{j \in Z^+}^{n \in Z} \quad \text{for} \quad L^2(Y)$$

we define $W: L^2(Y) \to L^2(X)$ by $g_0 \to f_0$, $U_S^n g_j \to U_T^n f_j$ and extend by linearity. Thus $WU_S = U_T W$ and S and T are spectrally isomorphic.

(2) Using a similar method to the one used above for the (½,½)-shift one can show that if at least two of $\{p_1, p_2, \ldots, p_n\}$ are non-zero then the 2-sided $\{p_1, p_2, \ldots, p_n\}$ -shift has countable Lebesgue spectrum.

Theorem 2.3:

If T has countable Lebesgue spectrum it is strong-mixing.

Proof: Let $\{f_0, U_T^n f_m : n \in Z, \ m > 0\}$ be the basis. Then, clearly, as $p \to \infty$

$$(U_T^p \circ U_T^n f_m, U_T^k f_q) \to (U_T^n f_m, 1)(1, U_T^k f_q) \quad \forall \ k, n \in Z, \quad m, q \geq 0.$$

Fix k and q and consider

$$H_{k,q} = \{f \in L^2(m): (U_T^p f, U_T^k f_q) \to (f, 1)(1, U_T^k f_q)\}.$$

$H_{k,q}$ is a closed subspace of $L^2(m)$ (c.f. proof in example (4) §6 Ch. 1) and contains the basis $\{f_0, U_T^n f_m : n \in Z, \ m > 0\}$ and hence is equal to $L^2(m)$. Fix $f \in L^2(m)$ and let $L_f = \{g \in L^2(m): (U_T^p f, g) \to (f, 1)(1, g)\}$. L_f is a closed subspace of $L^2(m)$, contains the basis by the above, and therefore is equal to the entirety of $L^2(m)$. Hence

$$(U_T^p f, g) \to (f, 1)(1, g) \quad \forall \ f, g \in L^2(m). \qquad //$$

Suppose A: G → G is an ergodic automorphism of a compact abelian metric group. Then the automorphism Â: Ĝ → Ĝ has no finite orbits except for the orbit of the identity. (This is what the ergodicity of A says.) Since the characters form a basis for $L^2(m)$ we can conclude that A has countable Lebesgue spectrum if we can show there are infinitely many distinct orbits of Â. This is proved in Halmos [2].

In Chapter 4 we shall consider a whole class of transformations with countable Lebesgue spectrum.

Chapter 3: Measure-Preserving Transformations

with Pure Point Spectrum

In this chapter we study a class of measure-preserving transformations for which the conjugacy problem is solved and for which spectral isomorphism implies conjugacy.

§1. Eigenfunctions

Suppose T is an ergodic measure-preserving transformation of a
probability space $(X, \mathcal{B}, m)$. Suppose λ is an eigenvalue corresponding to the eigenfunction f, i.e., $f \neq 0$, $f \in L^2(m)$, $U_T f =$
$\lambda f \in L^2(m)$ ($(f \circ T)(x) = \lambda f(x)$ a.e.). Then

(1) $|\lambda| = 1$ and $|f|$ is a constant a.e.

We have $f(T(x))\overline{f(T(x))} = \lambda \overline{\lambda} f(x)\overline{f(x)}$ a.e. Integrating both
sides we get that $\|f\|^2 = |\lambda|^2 \|f\|^2$. Therefore $|\lambda| = 1$. Also
$|f(T(x))| = |\lambda||f(x)|$ a.e. $= |f(x)|$ a.e. Thus $|f|$ is a T-invariant
function and, since T is ergodic, $|f| =$ a constant a.e.

(2) Eigenfunctions corresponding to different eigenvalues are
orthogonal.

Suppose $\lambda \neq \mu$, $U_T f = \lambda f$, $U_T g = \mu g$. Then

$$(f,g) = (U_T f, U_T g) = (\lambda f, \mu g) = \lambda \overline{\mu}(f,g)$$

and $\lambda \overline{\mu} \neq 1$ implies $(f,g) = 0$.

(3) If $f \circ T = \lambda f$, $g \circ T = \lambda g$ then $f = c \cdot g$ where c is some constant.

By (1) $g \neq 0$, so $(f/g) \circ T = f/g$ which must be constant since
T is ergodic.

So (2) and (3) show that eigenspaces are 1-dimensional and

mutually orthogonal.

(4) The eigenvalues of T form a subgroup of K.

   If $f \circ T = \lambda f$, $g \circ T = \mu g$ then $(f\bar{g}) \circ T = \lambda\bar{\mu}f\bar{g}$.

   By (2) if $L^2(m)$ is separable then the group of eigenvalues is countable.

§2. Pure Point Spectrum

Definition 3.1:

   An ergodic measure-preserving transformation T on a probability space (X, B, m) has pure point spectrum (discrete spectrum) if there exists an orthomormal basis for $L^2(m)$ which consists of eigenfunctions of T.

   The following theorem shows that the eigenvalues determine completely whether two such transformations are conjugate or not.

Theorem 3.1: (Discrete Spectrum Theorem - Halmos and Von Neumann
                [1], 1942)
   The following are equivalent for ergodic measure-preserving transformations T_1 and T_2 each having pure point spectrum:
(1) T_1 and T_2 are spectrally isomorphic.
(2) T_1 and T_2 have the same eigenvalues.
(3) T_1 and T_2 are conjugate.

   Proof: (1) ⇒ (2) is trivial.

   (3) ⇒ (1) is always true (see §3 of Chapter 2).

   (2) ⇒ (1). For each eigenvalue λ, choose $f_\lambda \in L^2(m_1)$, $g_\lambda \in L^2(m_2)$ such that

$$U_{T_1} f_\lambda = \lambda f_\lambda, \qquad U_{T_2} g_\lambda = \lambda g_\lambda$$

and

$$|f_\lambda| = |g_\lambda| = 1.$$

We define $W: L^2(m_2) \rightarrow L^2(m_1)$ by $W(g_\lambda) = f_\lambda$ and extending by linearity. We readily see that W is a bijective isometry; moreover $WU_{T_2} = U_{T_1}W$ by checking this on the g_λ.

(2) $\Rightarrow$ (3). To prove this we need the following result:

Theorem 3.2:

Let H be a discrete abelian group and K a divisible subgroup of H (i.e., $\forall$ k $\in$ K and $\forall$ n > 0 $\exists$ a $\in$ K $\ni$ $a^n = k$). Then there exists a homomorphism $\phi: H \rightarrow K$ such that $\phi|_K$ = identity (i.e., K is an algebraic retract of H).

Proof: Let R consist of all retracts onto K from supergroups of K in H, i.e., R consists of all pairs (M,ϕ) where $H \geq M \geq K$ and $\phi: M \rightarrow K$ is a homomorphism such that $\phi|_K$ = identity. R is non-empty as $(K,\text{id}_K) \in R$. We order R by extension, i.e., $(M_1,\phi_1) < (M_2,\phi_2)$ if $M_1 \leq M_2$ and $\phi_2|_{M_1} = \phi_1$. This is a partial ordering and every linearly ordered subset has an upper bound. So, by Zorn's Lemma there exists a maximal element, say (L,p), of R.

We claim that $L = H$. Suppose not, then consider $g \in H \setminus L$ and let M be the group generated by g and L.

Case 1: If no power of g lies in L then every element of M can be uniquely written in the form $g^i a$ where $a \in L$, $i \in Z$. We define $\psi: M \rightarrow K$ by $\psi(g^i a) = p(a)$. We can easily check that ψ is a homomorphism and that $\psi|_K = \text{id}_K$. This then contradicts the maximality of (L,p).

Case 2: Let n be the least positive integer such that $g^n \in L$. Each element of M can be uniquely written as $g^i a$, where $a \in L$, $0 \leq i \leq n-1$. Since K is divisible, let $g_0 \in K$ be such that $p(g^n) = g_0^n$. Then $\psi(g^i a) = g_0^i p(a)$ defines a homomorphism of M into K such that $\psi|_K = \text{id}_k$. Again, we have contradicted the maximality of (L,p).

Thus it follows that L = H. //

We now prove that (2) ⇒ (3). Let Λ denote the group of eigen-values of T_1 = the group of eigenvalues of T_2. Fix $\lambda \in \Lambda$. Let $f_\lambda \in L^2(m_1)$ be chosen so that $|f_\lambda| = 1$, $U_{T_1}f_\lambda = \lambda f_\lambda$ and observe that $\{f_\lambda : \lambda \in \Lambda\}$ is a basis for $L^2(m_1)$. Also, choose $g_\lambda \in L^2(m_2)$ so that $|g_\lambda| = 1$, $U_{T_2}g_\lambda = \lambda g_\lambda$ and observe that $\{g_\lambda : \lambda \in \Lambda\}$ is a basis for $L^2(m_2)$.

$$U_{T_i}f_{\lambda\mu} = \lambda\mu f_{\lambda\mu} \quad \forall \; \lambda,\mu \in \Lambda$$

and also

$$U_{T_i}(f_\lambda \cdot f_\mu) = f_\lambda(T) \cdot f_\mu(T) = (\lambda\mu)(f_\lambda \cdot f_\mu).$$

By (3) of §1 there exists a constant $r(\lambda,\mu) \in K$ such that $f_\lambda(x)f_\mu(x) = r(\lambda,\mu)f_{\lambda\mu}(x)$ a.e.

Let H denote the collection of all functions X → K. Clearly, H is an abelian group under pointwise multiplication. Moreover, K is a subgroup of H if we identify constant functions with their values.

By the previous Theorem 3.2 there exists a homomorphism $\phi: H \to K$ such that $\phi|_K = \mathrm{id}_K$. Let $f^* = \overline{\phi(f_\lambda)}f_\lambda$; then $|f_\lambda^*| = 1$, $U_T f_\lambda^* = \lambda f_\lambda^*$ and $\{f_\lambda^* : \lambda \in \Lambda\}$ is a basis for $L^2(m_1)$. Also,

$$f_\lambda^* f_\mu^* = \overline{\phi(f_\lambda)}\,\overline{\phi(f_\mu)}f_\lambda f_\mu = \overline{\phi(f_\lambda f_\mu)}f_\lambda f_\mu$$

$$= \overline{\phi(r(\lambda,\mu))}\,\overline{\phi(f_{\lambda\mu})}f_{\lambda\mu}r(\lambda,\mu)$$

$$= \overline{r(\lambda,\mu)}\,\overline{\phi(f_{\lambda\mu})}r(\lambda,\mu)f_{\lambda\mu}$$

$$= f_{\lambda\mu}^*.$$

Thus for all intents and purposes we can assume that $f_\lambda f_\mu = f_{\lambda\mu}$ and $g_\lambda g_\mu = g_{\lambda\mu}$.

Define $W: L^2(m_2) \to L^2(m_1)$ by $W(g_\lambda) = f_\lambda$ and extend by linearity. W is bijective, linear and preserves the inner product. Also, $WU_{T_2} = U_{T_1} W$. If we can show that W is multiplicative then W is necessarily induced by an isomorphism of measure algebras (by Corollary 2.1) and hence T_1 and T_2 are conjugate. But,

$$W(g_\lambda g_\mu) = W(g_{\lambda\mu}) = f_{\lambda\mu} = f_\lambda f_\mu = W(g_\lambda)W(g_\mu).$$

Let h,k be bounded functions in $L^2(m_2)$. If we fix g_μ and let a finite linear combination of g_μ's converge to h in $L^2(m_2)$ we obtain that $W(hg_\mu) = W(h)W(g_\mu)$. Then if we let a finite linear combination of g_μ's converge to k in $L^2(m_2)$ we get that $W(hk) = W(h)W(k)$. It follows from this that W maps bounded functions to bounded functions since this is also true for bounded h and any $k \in L^2(m_2)$, and then $W(h)$ is bounded since $W(h)f \in L^2(m_1)$ for all f in $L^2(m_1)$. //

<u>Corollary</u> 3.3:

If T is an invertible ergodic transformation with pure point spectrum then T and T^{-1} are conjugate.

<u>Proof</u>: They have the same eigenvalues. //

§3. <u>Group</u> <u>Rotations</u>

<u>Example</u>:

Let $T: K \to K$ be defined by $T(z) = az$ where a is not a root of unity. We know that T is ergodic. Let $f_n: K \to C$ be defined by $f_n(z) = z^n$ where $n \in Z$.

$$f_n(Tz) = f_n(az) = a^n z^n = a^n f_n(z).$$

Thus f_n is an eigenfunction with eigenvalue a^n. Since the $\{f_n\}$ form a basis for $L^2(K)$ we see that T is ergodic and has pure point spectrum.

These ideas carry over to rotations on compact abelian groups.

Theorem 3.3:

Let T, (T(g) = ag) be an ergodic rotation of a compact abelian group G. Then T has pure point spectrum. The eigenfunctions of T all consist of constant multiples of characters, and the eigenvalues are $\{\gamma(a): \gamma \in \hat{G}\}$.

Proof: Let $\gamma \in \hat{G}$. Then

$$\gamma(Tg) = \gamma(ag) = \gamma(a)\gamma(g).$$

Therefore each character is an eigenfunction and so T has pure point spectrum. If there is another eigenvalue besides the members of $\{\gamma(a): \gamma \in \hat{G}\}$ then the corresponding eigenfunction would be orthogonal to all members of $\hat{G}$, by (4) of §1, and so is zero. Hence $\{\gamma(a): \gamma \in \hat{G}\}$ consists of all the eigenvalues of T and the only eigenfunctions are constant multiples of characters, using (3) of §1. //

Theorem 3.4: (Representation Theorem)

An ergodic measure-preserving transformation T with pure point spectrum is conjugate to an ergodic rotation on some compact abelian group.

Proof: Let Λ = the group of all eigenvalues for T and give Λ the discrete topology. If $L^2(m)$ is separable then Λ is countable. In the other case we shall need to use the character theory of groups without a countable basis. Let $G = \hat{\Lambda}$, the character group of Λ. G is compact and abelian. The map $\alpha: \Lambda \to K$ given by $\alpha(\lambda) = \lambda$ is a homomorphism of the discrete group Λ and hence, by (2) of §5 of Chapter 0, $\exists$ a $\in G$ so that $\alpha(\lambda) = \underset{\sim}{\lambda}(a)$ $\lambda \in \Lambda$ (where we write $\underset{\sim}{\lambda}$ when we wish to consider "λ" as a homomorphism of G to K).

Define $S: G \to G$ by $S(g) = ag$. We claim that S is ergodic. Suppose $fS = f$, $f \in L^2(G)$, $f = \sum_j b_j \underset{\sim}{\lambda}_j$, $\lambda_j \in \Lambda$. Then the above gives us that

$$\sum_j b_j \underset{\sim}{\lambda}_j(a) \underset{\sim}{\lambda}_j(g) \quad \sim \quad \sum_j b_j \underset{\sim}{\lambda}_j(g)$$

so, $b_j \underset{\sim}{\lambda}_j(a) = b_j$. But $\underset{\sim}{\lambda}_j(a) = \lambda_j$ and therefore $b_j \lambda_j = b_j$. If $b_j \neq 0$ then necessarily $\lambda_j = 1$. Thus $\underset{\sim}{\lambda}_j(g) = 1$ for all $g \in G$, and we get that $f = $ a constant a.e. We know then that S is ergodic, and by the previous theorem has pure point spectrum.

Again by the previous theorem the eigenvalues of $S = \{\gamma(a): \gamma \in \hat{G}\} = \{\alpha(\lambda): \lambda \in \Lambda\} = \{\lambda: \lambda \in \Lambda\} = \Lambda$. So, S and T have the same eigenvalues and both have pure point spectrum. Hence the Discrete Spectrum Theorem tells us that they are conjugate. //

Theorem 3.5: (Existence Theorem)

Every subgroup Λ of K is the group of eigenvalues of an ergodic measure-preserving transformation with pure point spectrum.

Proof: The desired transformation is the rotation S constructed in the proof of Theorem 3.4. //

The conjugacy problem for ergodic measure-preserving transformations with pure point spectrum is completely solved. We have some very simple invariants, namely the eigenvalues, which determine when two such transformations are conjugate. Also there are some simple examples, namely group rotations, such that each ergodic measure-preserving transformation with pure point spectrum is conjugate to one of these examples. So each conjugacy class of ergodic measure-preserving transformations with pure point spectrum is characterized by a subgroup of K, and each subgroup of K corresponds to a conjugacy class.

Chapter 4: Entropy

We are searching for conjugacy and/or isomorphism invariants.
In 1958 Kolmogorov [1] introduced the concept of entropy into ergodic
theory, and this has been the most successful invariant so far. For
example, in 1943 it was known that the two-sided (1/2,1/2)-shift and
the two-sided (1/3,1/3)-shift both have countable Lebesgue spectrum
and hence are spectrally isomorphic; but it was not known whether they
were conjugate. This was resolved in 1958 when Kolmogorov showed that
they had entropies log 2 and log 3 respectively and hence are not
conjugate. Von Neumann had had the same idea considerably earlier,
but he was unable to prove that entropy was a conjugacy invariant.
The notion of entropy now used is slightly different from that used
by Kolmogorov - the improvement was made by Sinai [1] in 1959.

§1. Partitions and Subalgebras

Throughout, (X,B,m) will denote a probability space.

Definition 4.1:

A partition of (X,B,m) is a disjoint collection of elements
of B whose union is X.

We shall be interested in finite partitions. They will be denoted by Greek letters, e.g., $\xi = \{A_1,\ldots,A_k\}$.

If ξ is a finite partition of (X,B,m) then the collection of
all elements of B which are unions of elements of ξ is a finite
sub-σ-algebra of B. We denote it by $A(\xi)$. Conversely, if C is a
finite sub-σ-algebra of B, say $C = \{C_i : i = 1,\ldots,n\}$, then the non-
empty sets of the form $B_1 \cap \ldots \cap B_n$ where $B_i = C_i$ or $X \setminus C_i$ form
a finite partition of (X,B,m). We denote it by $\xi(C)$. Thus we have

a one-to-one correspondence between finite partitions and finite sub-σ-algebras of B.

Definition 4.2:

Suppose ξ and η are two finite partitions. $\xi \le \eta$ means that each element of ξ is a union of elements of η.

Note:
$$\xi \le \eta \iff A(\xi) \subseteq A(\eta)$$
$$A \subseteq C \iff \xi(A) \le \xi(C).$$

Definition 4.3:

Let $\xi = \{A_1, \ldots, A_n\}$, $\eta = \{C_1, \ldots, C_k\}$. Then

$$\xi \vee \eta = \{A_i \cap C_j : 1 \le i \le n, \quad 1 \le j \le k\}.$$

If A and C are finite sub-σ-algebras of B then $A \vee C$ denotes the smallest sub-σ-algebra of B containing A and C.

Note:
$$\xi(A \vee C) = \xi(A) \vee \xi(C)$$
$$A(\xi \vee \eta) = A(\xi) \vee A(\eta).$$

Suppose $T: X \to X$ is a measure-preserving transformation. If $\xi = \{A_1, \ldots, A_m\}$, then by $T^{-n}\xi$ we mean $\{T^{-n}A_1, \ldots, T^{-n}A_m\}$ and by $T^{-n}(A)$ we mean $\{T^{-n}A : A \in A\}$ $(n > 0)$.

Note: If $n \ge 0$
$$\xi(T^{-n}A) = T^{-n}\xi(A)$$
$$A(T^{-n}\xi) = T^{-n}A(\xi)$$
$$T^{-n}(A \vee C) = T^{-n}A \vee T^{-n}C$$
$$T^{-n}(\xi \vee \eta) = T^{-n}\xi \vee T^{-n}\eta$$
$$\xi \le \eta \implies T^{-n}\xi \le T^{-n}\eta$$
$$A \subseteq C \implies T^{-n}A \subseteq T^{-n}C.$$

Definition 4.4:

If $\mathcal{D}$ and $\mathcal{E}$ are (not necessarily finite) sub-σ-algebras of $\mathcal{B}$, then we write $\mathcal{D} \overset{\circ}{=} \mathcal{E}$ if $\forall$ $D \in \mathcal{D}$ $\exists$ $E \in \mathcal{E}$ such that $m(D \Delta E) = 0$ and $\forall$ $E \in \mathcal{E}$ $\exists$ $D \in \mathcal{D}$ such that $m(D \Delta E) = 0$.

If $\mathcal{D}$ and $\mathcal{E}$ are finite, $\mathcal{D} \overset{\circ}{=} E$, and if $\xi(\mathcal{D}) = \{D_1, \ldots, D_p, D_{p+1}, \ldots, D_q\}$ where $m(D_i) > 0$ for $1 \leq i \leq p$ and $m(D_i) = 0$ for $p+1 \leq i \leq q$, then $\xi(E) = \{E_1, \ldots, E_p, E_{p+1}, \ldots, E_s\}$ where $m(E_i \Delta D_i) = 0$ for $1 \leq i \leq p$ and $m(E_i) = 0$ for $p+1 \leq i \leq s$.

$\mathcal{D} \overset{\circ}{\subset} \mathcal{C}$ means $\forall$ $D \in \mathcal{D}$ $\exists$ $C \in \mathcal{C}$ such that $m(D \Delta C) = 0$.

§2. Entropy

All logarithms are to base 2 and $0 \cdot \log 0 = 0$.

Let $A \subseteq \mathcal{B}$ be finite. Let $\xi(A) = \{A_1, \ldots, A_k\}$. Then

$$H(A) = H(\xi(A)) = - \sum_{i=1}^{k} m(A_i) \log m(A_i),$$

is called the __entropy__ of __A__ (or of $\xi(A)$). (This means that if $A_1, \ldots, A_k$ denote the outcomes of an experiment then $H(A)$ measures the uncertainty removed (or information gained) by performing the experiment. $H(A)$ is a measure of the uncertainty about which A_i a general point of X will belong to.)

Remarks:

(1) If $A = \{X, \phi\}$ then $H(A) = 0$. Here A represents the outcomes of a "certain" experiment so there is no uncertainty about the outcome.

(2) If $\xi(A) = \{A_1, \ldots, A_k\}$ where $m(A_i) = 1/k$ $\forall$ i then

$$H(A) = - \sum_{i=1}^{k} \frac{1}{k} \log \frac{1}{k} = \log k.$$

Thus, we gain a lot of information if k is large. (Since all the members of ξ(A) have equal measure there is much uncertainty about which A_i a point will belong to.)

(3) H(A) ≥ 0.

(4) If A $\overset{o}{=}$ C then H(A) = H(C).

Suppose T: X → X is measure-preserving.

If A is a finite sub-σ-algebra of B we define

$$h(T,A) = \lim_{n \to \infty} \frac{1}{n} H(A \vee T^{-1}A \vee \dots \vee T^{-(n-1)}A)$$

$$= \lim_{n \to \infty} \frac{1}{n} H\left(\bigvee_{i=0}^{n-1} T^{-i}A \right),$$

which we call the <u>entropy</u> <u>of</u> <u>T</u> <u>with</u> <u>respect</u> <u>to</u> <u>A</u>. (Later (in Corollary 4.4) we will show that the above limit always exists.) (This means that if we think of an application of T as a passage of one day of time, then $\bigvee_{i=0}^{n-1} T^{-i}A$ represents the combined experiment of performing the original experiment represented by A on n consecutive days. h(T,A) is then the average information per day that one gets from performing the original experiment daily forever.)

<u>Remarks</u>:

(5) h(T,A) ≥ 0.

(6) The elements of $\xi\left(\bigvee_{i=0}^{n-1} T^{-i}A \right) = \bigvee_{i=0}^{n-1} \xi(T^{-i}A)$ are all the sets of the form $\bigcap_{i=0}^{n-1} T^{-i}A_{m_i}$ where ξ(A) = {$A_1, \dots, A_k$}.

We define h(T) = sup h(T,A) where the supremum is taken over all finite sub-σ-algebras A contained in B and call this the <u>entropy</u> <u>of</u> <u>T</u>. (h(T) is the maximum average information per day obtainable by performing a finite experiment.)

Remarks:

(7) $h(T) \geq 0$. $h(T)$ could be $+\infty$.

(8) $h(\mathrm{id}_X) = 0$. If $h(T) = 0$ then $h(T,A) = 0$ for every finite A,
which implies that $\bigvee_{i=0}^{n-1} T^{-i}A$ does not change much as $n \to \infty$.

Theorem 4.1:

Entropy is a conjugacy invariant and hence an isomorphism invariant.

Proof: Let $T_1: X_1 \to X_1$, $T_2: X_2 \to X_2$ be measure-preserving
and let $\Phi: (\mathcal{B}_2, m_2) \to (\mathcal{B}_1, m_1)$ be an isomorphism of measure algebras
such that $\Phi T_2^{-1} = T_1^{-1}\Phi$. Let A_2 be finite, $A_2 \subset \mathcal{B}_2$, and $\xi(A_2) =$
$\{A_1,\ldots,A_r\}$. Choose $B_i \in \mathcal{B}_1$ such that $B_i = \Phi(A_i)$ and so that
$\eta = \{B_1,\ldots,B_r\}$ forms a partition of (X_1,B_1,m_1). Let $A_1 = A(\eta)$.

Now $\bigcap_{i=0}^{n-1} T_1^{-i}B_{q_i}$ (where the $q_i \in \{1,\ldots,r\}$) has the same measure

as $\bigcap_{i=0}^{n-1} T_2^{-i}A_{q_i}$ since

$$\Phi(\bigcap_{i=0}^{n-1} T_2^{-i}A_{q_i}) = \Phi(\bigcap_{i=0}^{n-1} T_2^{-i}A_{q_i}) = \bigcap_{i=0}^{n-1} T_1^{-i}\Phi(A_{q_i}) = \bigcap_{i=0}^{n-1} T_1^{-i}B_{q_i} = \bigcap_{i=0}^{n-1} T_1^{-i}B_{q_i}.$$

Thus, $H(\bigvee_{i=0}^{n-1} T_1^{-i}A_1) = H(\bigvee_{i=0}^{n-1} T_2^{-i}A_2)$ which implies that $h(T_1,A_1) =$
$h(T_2,A_2)$ which in turn implies $h(T_1) \geq h(T_2)$. By symmetry we then
get that $h(T_1) = h(T_2)$. //

Theorem 4.2:

The function $\phi: [0,\infty) \to R$ defined by:

$$\phi(x) = \begin{cases} 0 & \text{if } x = 0 \\ x \cdot \log x & \text{if } x \neq 0 \end{cases}$$

is convex, i.e., $\phi(\alpha x + \beta y) \leq \alpha\phi(x) + \beta\phi(y)$ if $x,y \in [0,\infty)$, $\alpha,\beta \geq 0$,
$\alpha+\beta = 1$.

By induction

$$\phi(\sum_{i=1}^{k} \alpha_i x_i) \leq \sum_{i=1}^{k} \alpha_i \phi(x_i)$$

if $x_i \in [0,\infty)$, $\alpha_i \geq 0$, $\sum_{i=1}^{k} \alpha_i = 1$.

Proof:

$$\phi'(x) = \log e + \log x$$

$$\phi''(x) = \frac{\log e}{x} > 0 \quad \text{on} \quad (0,\infty).$$

Suppose $y > x$; by the mean value theorem

$$\phi(y) - \phi(\alpha x + \beta y) = \phi'(z)\alpha(y-x)$$

$$\text{where} \quad \alpha x + \beta y < z < y \quad \text{and}$$

$$\phi(\alpha x + \beta y) - \phi(x) = \phi'(w)\beta(y-x)$$

$$\text{where} \quad x < w < \alpha x + \beta y.$$

Since $\phi'' > 0$ $\phi'(z) \geq \phi'(w)$, thus

$$\beta(\phi(y) - \phi(\alpha x + \beta y)) = \phi'(z)\alpha\beta(y-x)$$

$$\geq \phi'(w)\alpha\beta(y-x) = \alpha(\phi(x+\beta y) - \phi(x)).$$

Therefore $\phi(\alpha x + \beta y) \leq \alpha\phi(x) + \beta\phi(y)$ if $x,y > 0$, and hence also if $x,y \geq 0$ by continuity of ϕ. //

Corollary 4.2:

If $\xi = \{A_1,\ldots,A_k\}$ then $H(\xi) \leq \log k$.

Proof: Put $\alpha_i = 1/k$ and $x_i = m(A_i)$ $1 \leq i \leq k$. Then $H(\xi) \leq \log k$. //

Combined with remark (2) this corollary shows that among all the partitions of X into k sets, the largest entropy is obtained when all the sets have equal measure. This fits in with our intuitive interpretation of entropy.

§3. Conditional Entropy

Let $A, C \subseteq B$ be finite.

$$\xi(A) = \{A_1, \ldots, A_k\}, \qquad \xi(C) = \{C_1, \ldots, C_p\}.$$

We define the <u>entropy</u> <u>of</u> A <u>given</u> C to be

$$H(A/C) = - \sum_{j=1}^{p} m(C_j) \sum_{i=1}^{k} \frac{m(A_i \cap C_j)}{m(C_j)} \log \frac{m(A_i \cap C_j)}{m(C_j)}$$

$$= - \sum_{i,j} m(A_i \cap C_j) \log \frac{m(A_i \cap C_j)}{m(C_j)} \geq 0$$

omitting the j-terms when $m(C_j) = 0$.

So to get $H(A/C)$ one considers C_j as a measure space with normalized measure $m(\cdot)/m(C_j)$ and calculates the entropy of the partition of C_j induced by $\xi(A)$ (this gives

$$- \sum_{i=1}^{k} \frac{m(A_i \cap C_j)}{m(C_j)} \log \frac{m(A_i \cap C_j)}{m(C_j)})$$ and then averages the answer taking

into account the size of C_j. ($H(A/C)$ measures the average information obtained from performing the experiment associated with A given the outcome of the experiment associated with C.)

Let N denote the σ-field $\{\phi, X\}$. Then $H(A/N) = H(A)$. (Since N represents the outcome of the trivial experiment one gains nothing from knowledge of it.)

Remarks:

(1) $H(A/C) \geq 0$.

(2) If $A \cong D$ then $H(A/C) = H(D/C)$.

(3) If $C \cong D$ then $H(A/C) = H(A/D)$.

Theorem 4.3:

If A, C, D are finite subalgebras of B then:

(i) $H(A \vee C/D) = H(A/D) + H(C/A \vee D)$

(ii) $H(A \vee C) = H(A) + H(C/A)$

(iii) $A \subseteq C \Rightarrow H(A/D) \leq H(C/D)$

(iv) $A \subseteq C \Rightarrow H(A) \leq H(C)$

(v) $C \subseteq D \Rightarrow H(A/C) \geq H(A/D)$

(vi) $H(A) \geq H(A/D)$

(vii) $H(A \vee C/D) \leq H(A/D) + H(C/D)$

(viii) $H(A \vee C) \leq H(A) + H(C)$.

(ix)   If T is measure-preserving then:

                       $H(T^{-1}A/T^{-1}C) = H(A/C)$ and

(x) $H(T^{-1}A) = H(A)$.

(The reader should think of the intuitive meaning of each statement.
This enables one to remember these results easily.)

   Proof: Let $\xi(A) = \{A_i\}$, $\xi(C) = \{C_j\}$, $\xi(D) = \{D_k\}$ and assume,
without loss of generality, that all sets have strictly positive
measure (since if $\xi(A) = \{A_1, \ldots, A_k\}$ with $m(A_i) > 0$ $1 \leq i \leq r$
and $m(A_i) = 0$ $r \leq i \leq k$ we can replace $\xi(A)$ by
 $\{A_1, \ldots, A_{r-1}, A_r \cup A_{r+1} \cup \ldots \cup A_k\}$).

(i) $H(A \vee C/D) = - \sum_{i,j,k} m(A_i \cap C_j \cap D_k) \log \dfrac{m(A_i \cap C_j \cap D_k)}{m(D_k)}$.

But $\dfrac{m(A_i \cap C_j \cap D_k)}{m(D_k)} = \dfrac{m(A_i \cap C_j \cap D_k)}{m(A_i \cap D_k)} \dfrac{m(A_i \cap D_k)}{m(D_k)}$ unless $m(A_i \cap D_k) = 0$

and then the left hand side is zero and we need not consider it; and
therefore

$$H(A \vee C/D) = - \sum_{i,j,k} m(A_i \cap C_j \cap D_k) \log \frac{m(A_i \cap D_k)}{m(D_k)}$$

$$- \sum_{i,j,k} m(A_i \cap C_j \cap D_k) \log \frac{m(A_i \cap C_j \cap D_k)}{m(A_i \cap D_k)}$$

$$= - \sum_{i,k} m(A_i \cap D_k) \log \frac{m(A_i \cap D_k)}{m(D_k)} + H(C/A \vee D)$$

$$= H(A/D) + H(C/A \vee D).$$

(ii)   Put $D = N = \{\phi, X\}$.

(iii)   By (i)

$$H(C/D) = H(A \vee C/D) = H(A/D) + H(C/A \vee D) \geq H(A/D).$$

(iv)   Put $D = N$ in (iii).

(v)   Fix i,j and let $\alpha_k = \dfrac{m(D_k \cap C_j)}{m(C_j)}$, $x_k = \dfrac{m(A_i \cap D_k)}{m(D_k)}$.   Then by Theorem 4.2

$$\phi \left(\sum_k \frac{m(D_k \cap C_j)}{m(C_j)} \frac{m(A_i \cap D_k)}{m(D_k)} \right) \leq \sum_k \frac{m(D_k \cap C_j)}{m(C_j)} \phi \left(\frac{m(A_i \cap D_k)}{m(D_k)} \right)$$

but since $C \subseteq D$ the left hand side

$$= \phi \left(\frac{m(A_i \cap C_j)}{m(C_j)} \right) = \frac{m(A_i \cap C_j)}{m(C_j)} \log \frac{m(A_i \cap C_j)}{m(C_j)}.$$

Multiply both sides by $m(C_j)$ and sum over i and j to give

$$\sum_{i,j} m(A_i \cap C_j) \log \frac{m(A_i \cap C_j)}{m(C_j)} \leq \sum_{i,j,k} m(D_k \cap C_j) \frac{m(A_i \cap D_k)}{m(D_k)} \log \frac{m(A_i \cap D_k)}{m(D_k)}$$

$$= \sum_{i,k} m(D_k) \frac{m(A_i \cap D_k)}{m(D_k)} \log \frac{m(A_i \cap D_k)}{m(D_k)}$$

or $\qquad$ $-H(A/C) \leq -H(A/D).$

Therefore $\qquad$ $H(A/D) \leq H(A/C).$

(vi)   Put $C = N$ in (v).

(vii)   Use (i) and (v).

(viii) Set $D = N$ in (vii).

(ix), (x) Clear from definitions. //

Theorem 4.4:

If $\{a_n\}_{n \geq 1}$ satisfies $a_n \geq 0$, $a_{n+m} \leq a_n + a_m$ $\forall$ n,m, then $\lim\limits_{n \to \infty} a_n/n$ exists and equals $\inf\limits_n a_n/n$.

Proof: Fix $m > 0$. For each $j > 0$ $j = km + n$ where $0 \leq n < m$. Then

$$\frac{a_j}{j} = \frac{a_{n+km}}{n+km} \leq \frac{a_n}{km} + \frac{a_{km}}{km} \leq \frac{a_n}{km} + \frac{ka_m}{km} = \frac{a_n}{km} + \frac{a_m}{m} .$$

As $j \to \infty$ then $k \to \infty$ so $\overline{\lim} \dfrac{a_j}{j} \leq \dfrac{a_m}{m}$ and therefore $\overline{\lim} \dfrac{a_j}{j} \leq \inf \dfrac{a_m}{m}$. But $\inf \dfrac{a_m}{m} \leq \underline{\lim} \dfrac{a_j}{j}$ so that $\lim \dfrac{a_j}{j}$ exists and equals $\inf \dfrac{a_j}{j}$. //

Corollary 4.4:

If $A \subseteq B$ then $\lim\limits_{n \to \infty} \dfrac{1}{n} H(\bigvee\limits_{i=0}^{n-1} T^{-i}A)$ exists.

(See also the remark after Theorem 4.6.)

Proof: Let $a_n = H(\bigvee\limits_{i=0}^{n-1} T^{-i}A) \geq 0.$

$$a_{n+m} = H(\bigvee_{i=0}^{n+m-1} T^{-i}A)$$

$$\leq H(\bigvee_{i=0}^{n-1} T^{-i}A) + H(\bigvee_{i=n}^{n+m-1} T^{-i}A) \qquad \begin{array}{l} \text{by (viii)} \\ \text{of Theorem 4.3.} \end{array}$$

$$= a_n + H(\bigvee_{i=0}^{m-1} T^{-i}A) \quad \text{by (x) in Theorem 4.3.}$$

$$= a_n + a_m.$$

We then apply Theorem 4.4. //

§4. Properties of h(T,A)

Recall that $h(T,A) = \lim\limits_{n\to\infty} \dfrac{1}{n} H(\bigvee\limits_{i=0}^{n-1} T^{-i}A)$.

Theorem 4.5:

Suppose A, C are subalgebras of B and T is measure-preserving. Then

(1) $\qquad\qquad\qquad h(T,A) \leq H(A).$

(2) $\qquad\qquad\qquad h(T,A \vee C) \leq h(T,A) + h(T,C).$

(3) $\qquad\qquad\qquad A \subseteq C \Rightarrow h(T,A) \leq h(T,C).$

(4) $\qquad\qquad\qquad h(T,A) \leq h(T,C) + H(A/C).$

(5) If T is invertible and $m \geq 1$ then

$$h(T,A) = h(T, \bigvee_{i=-m}^{m} T^i A).$$

Proof:

(1) $\qquad \dfrac{1}{n} H(\bigvee\limits_{i=0}^{n-1} T^{-i}A) \leq \dfrac{1}{n} \sum\limits_{i=0}^{n-1} H(T^{-i}A) \quad$ by (viii) of Theorem 4.3.

$$= \dfrac{1}{n} \sum_{i=0}^{n-1} H(A) \quad \text{by (x) of Theorem 4.3.}$$

$$= H(A).$$

(2) $\quad H(\bigvee_{i=0}^{n-1} T^{-i}(A \vee C)) = H(\bigvee_{i=0}^{n-1} T^{-i}A \vee \bigvee_{i=0}^{n-1} T^{-i}C)$

$\qquad\qquad \leq H(\bigvee_{i=0}^{n-1} T^{-i}A) + H(\bigvee_{i=0}^{n-1} T^{-i}C) \quad$ by (viii) of Theorem 4.3.

(3) $\quad$ If $\quad A \subseteq C \quad$ then

$$\bigvee_{i=0}^{n-1} T^{-i}A \subseteq \bigvee_{i=0}^{n-1} T^{-i}C \quad \forall n \geq 1$$

so one uses (iv) of Theorem 4.3.

(4) $\quad H(\bigvee_{i=0}^{n-1} T^{-i}A) \leq H((\bigvee_{i=0}^{n-1} T^{-i}A) \vee (\bigvee_{i=0}^{n-1} T^{-i}C)) \quad$ by (iv) of Theorem 4.3

$\qquad = H(\bigvee_{i=0}^{n-1} T^{-1}C) + H((\bigvee_{i=0}^{n-1} T^{-1}A)/(\bigvee_{i=0}^{n-1} T^{-1}C)) \,$ by (ii) of Theorem 4.3.

But, by (vii) of Theorem 4.3

$$H((\bigvee_{i=0}^{n-1} T^{-i}A)/(\bigvee_{i=0}^{n-1} T^{-i}C)) \leq \sum_{i=0}^{n-1} H(T^{-i}A/(\bigvee_{j=0}^{n-1} T^{-j}C))$$

$$\leq \sum_{i=0}^{n-1} H(T^{-i}A/T^{-i}C) \quad \text{by (v) of Theorem 4.3}$$

$$= nH(A/C) \quad \text{by (ix) of Theorem 4.3.}$$

Thus, $\qquad\qquad H(\bigvee_{i=0}^{n-1} T^{-i}A) \leq H(\bigvee_{i=0}^{n-1} T^{-i}C) + nH(A/C).$

(5) $\qquad\qquad h(T , \bigvee_{-m}^{m} T^{i}A) = \lim_{k \to \infty} \frac{1}{k} H(\bigvee_{j=0}^{k-1} T^{-j}(\bigvee_{i=-m}^{m} T^{i}A))$

$$= \lim_{k \to \infty} \frac{1}{k} H(\bigvee_{-m}^{m+k-1} T^{-i}A).$$

$$H(\bigvee_{-m}^{m+k-1} T^{-i}A) = H(\bigvee_{-m}^{m} T^{-i}A \vee \bigvee_{m}^{m+k-1} T^{-i}A)$$

$$= H(\bigvee_{m}^{m+k-1} T^{-i}A) + H((\bigvee_{-m}^{m} T^{-i}A)/(\bigvee_{m}^{m+k-1} T^{-i}A))$$

by (ii) of Theorem 4.3

$$= H(\bigvee_{i=0}^{k-1} T^{-i}A) + H((\bigvee_{-m}^{m} T^{-i}A)/(\bigvee_{m}^{m+k-1} T^{-i}A))$$

by (x) of Theorem 4.3.

We want to show that

$$\frac{1}{k} H((\bigvee_{-m}^{m} T^{-i}A)/(\bigvee_{m}^{m+k-1} T^{-i}A)) \to 0 \text{ as } k \to \infty.$$

But, $\frac{1}{k} H((\bigvee_{-m}^{m} T^{-i}A)/(\bigvee_{m}^{m+k-1} T^{-i}A)) \leq \frac{1}{k} H(\bigvee_{-m}^{m} T^{-i}A)$ by (vi) of Theorem 4.3 and so the result follows. //

Theorem 4.6:

If $A \subseteq B$ is finite, and T is measure-preserving then

$$h(T,A) = \lim_{n\to\infty} H(A/(\bigvee_{i=1}^{n} T^{-i}A)).$$

Proof: The limit exists since the right hand side is non-increasing in n by virtue of (v) of Theorem 4.3. We show by induction that for $n \geq 2$

$$H(\bigvee_{i=0}^{n-1} T^{-i}A) = H(A) + \sum_{j=1}^{n-1} H(A/(\bigvee_{i=1}^{j} T^{-i}A)).$$

This is true for $n = 2$ by (ii) of Theorem 4.3. Assume that this equality is true for n. We shall prove that it holds for $n+1$.

$$H(\bigvee_{i=0}^{n} T^{-i}A) = H(\bigvee_{i=1}^{n} T^{-i}A) + H(A/(\bigvee_{i=1}^{n} T^{-i}A))$$

by (ii) of Theorem 4.3

$$= H(\bigvee_{i=0}^{n-1} T^{-i}A) + H(A/(\bigvee_{i=1}^{n} T^{-i}A))$$

by (x) of Theorem 4.3

$$= H(A) + \sum_{j=1}^{n} H(A/(\bigvee_{i=1}^{j} T^{-i}A)). \qquad (*)$$

Dividing the above by n and taking the limit the result follows, using the fact that the Cesaro limit of a convergent sequence is the ordinary limit. //

Remark:

$$\frac{1}{n} H(\bigvee_{i=0}^{n-1} T^{-i}A) \text{ decreases to } h(T,A).$$

<u>Proof</u>: By (*) $\quad H(\bigvee_{i=0}^{n-1} T^{-i}A) \geq nH(A/(\bigvee_{i=1}^{n} T^{-i}A))$ using (v) of

Theorem 4.3. Hence

$$n[H(\bigvee_{i=0}^{n-1} T^{-i}A) + H(A/(\bigvee_{i=1}^{n} T^{-i}A))] \leq (n+1)H(\bigvee_{i=0}^{n-1} T^{-i}A)$$

i.e., $\qquad nH(\bigvee_{i=0}^{n} T^{-i}A) \leq (n+1)H(\bigvee_{i=0}^{n-1} T^{-i}A).$

§5. Properties of h(T)

<u>Theorem 4.7</u>:

(1) For $m > 0$, $h(T^m) = mh(T)$.

(2) If T is invertible then $h(T^m) = |m|h(T) \; \forall \, m \in Z$.

<u>Proof</u>: We first show that

$$h(T^m, \bigvee_{i=0}^{m-1} T^{-i}A) = mh(T,A).$$

This follows since

$$\lim_{k\to\infty} \frac{1}{k} H(\bigvee_{j=0}^{k-1} T^{-mj}(\bigvee_{i=0}^{m-1} T^{-i}A)) = \lim_{k\to\infty} \frac{m}{km} H(\bigvee_{i=0}^{km-1} T^{-i}A)$$

$$= mh(T,A).$$

Thus, $\quad mh(T) = m \cdot \sup_{A \text{ finite}} h(T,A) = \sup_A h(T^m, \bigvee_{i=0}^{m-1} T^{-i}A)$

$$\leq \sup_C h(T^m, C) = h(T^m).$$

Also, $h(T^m,A) \leq h(T^m, \bigvee_{i=0}^{m-1} T^{-i}A) = mh(T,A)$ by (3) of Theorem 4.5 and

so, $h(T^m) \leq mh(T)$. The result follows from these two inequalities.

(2) It suffices to show that $h(T^{-1}) = h(T)$; and all we need to show
is that $h(T^{-1},A) = h(T,A)$ for all finite A. But

$$H(\bigvee_{i=0}^{n-1} T^iA) = H(T^{-(n-1)} \bigvee_{i=0}^{n-1} T^iA) \quad \text{by (x) of Theorem 4.3}$$

$$= H(\bigvee_{j=0}^{n-1} T^{-j}A). \quad //$$

Theorem 4.8:

Let (X,B,m) be a probability space and B_0 be an algebra such
that the σ-algebra generated by B_0 (denoted by $\sigma(B_0)$) satisfies
$\sigma(B_0) \stackrel{\circ}{=} B$. Let C be a finite subalgebra of B. Then for every
$\varepsilon > 0$, there exists a finite algebra D, $D \subseteq B_0$ such that
$H(D/C) + H(C/D) < \varepsilon$.

Proof: Let $\xi(C) = \{C_1,\dots,C_r\}$ and assume without loss of
generality that each C_i has positive measure. We can do this since
if $C_1,\dots,C_s$ have positive measure and $C_{s+1},\dots,C_r$ have zero
measure where $1 \leq s \leq r$ then $\xi(C') = \{C_1,\dots,C_{s-1},C_s \cup \dots \cup C_r\}$
is such that $H(C'/D) = H(C/D)$ and $H(D/C') = H(D/C)$.

Since $\phi(x) = x \cdot \log x$ is continuous and $\phi(0) = 0$, $\phi(1) = 0$,

$\exists$ $0 < \delta_0 < 1$ such that $-\phi(x) < \varepsilon/2r$ if $0 \le x \le \delta_0$ or $1-\delta_0 \le x \le 1$.

We first show that if we choose a partition $\xi(\mathcal{D}) = \{D_1,\dots,D_r\}$ $\ni$ $m(C_i \Delta D_i) \le \min\limits_{1 \le i \le r} \delta_0 \dfrac{m(C_i)}{2} \equiv \delta$ $\forall$ i, then $H(C/\mathcal{D}) < \varepsilon/2$. For, if $\xi(\mathcal{D})$ satisfies the above inequality, then $m(C_i) < m(D_i) + \delta < m(D_i) + \dfrac{m(C_i)}{2}$ which implies $\dfrac{m(C_i)}{2} < m(D_i)$ and hence,

$$m(D_i) - m(D_i \cap C_i) \le m(D_i \Delta C_i) < \delta < \delta_0 m(D_i).$$

Thus $\qquad\qquad\qquad \dfrac{m(D_i \cap C_i)}{m(D_i)} \ge 1 - \delta_0.$

Therefore, if $j \ne i$ then $\dfrac{m(C_j \cap D_i)}{m(D_i)} \le \delta_0$ and hence,

$$H(C/\mathcal{D}) = -\sum_i m(D_i) \sum_j \phi\left(\frac{m(C_j \cap D_i)}{m(D_i)}\right)$$

$$\le \sum_i m(D_i) \sum_j \varepsilon/2r = \varepsilon/2.$$

So $m(C_i \Delta D_i) < \dfrac{\delta_0}{2} \min\limits_{1 \le i \le r} m(C_i)$ implies $H(C/\mathcal{D}) < \varepsilon/2$. If $m(C_i \Delta D_i) \le \dfrac{\delta_0}{4} \min\limits_{1 \le i \le r} m(C_i)$ then this still holds and also $m(C_i \Delta D_i) <$ $m(C_i)/4$ which implies $m(D_i) > \dfrac{3}{4} m(C_i)$ and hence

$$m(C_i \Delta D_i) < \frac{\delta_0}{4} \cdot \frac{4}{3} \min\limits_{1 \le i \le r} m(D_i) < \frac{\delta_0}{2} \min\limits_{1 \le i \le r} m(D_i).$$

Therefore $H(\mathcal{D}/C) < \varepsilon/2$ also.

So it suffices to show we can choose $D_i \in \mathcal{B}_0$ with

$$m(C_i \Delta D_i) \le \frac{\delta_0}{4} \min\limits_{1 \le i \le r} m(C_i) = \sigma.$$

Choose $\lambda > 0 \ni \lambda(r-1)[1 + r(r-1)] < \sigma$; for each i, choose
$B_i \in B_0 \ni m(C_i \Delta B_i) < \lambda$. If $i \neq j$ then $B_i \cap B_j \subset (B_i \Delta C_i) \cup (B_j \Delta C_j)$
so that $m(B_i \cap B_j) < 2\lambda$. Let $N = \bigcup_{i \neq j} (B_i \cap B_j)$. We have

$m(N) < r(r-1)\lambda$. Set $D_i = B_i \backslash N$ for $1 \leq i < r$ and $D_r = X \backslash \bigcup_{i=1}^{r-1} D_i$.

$\{D_1, \ldots, D_r\}$ is a partition of X and each $D_i \in B_0$. If $i < r$ then
$D_i \Delta C_i \subset (B_i \Delta C_i) \cup N$ and so,

$$m(D_i \Delta C_i) < \lambda[1 + r(r-1)] < \sigma.$$

However $D_r \Delta C_r \subset \bigcup_{i=1}^{r-1} (D_i \Delta C_i)$ and therefore

$$m(D_r \Delta C_r) < (r-1)\lambda[1 + r(r-1)] < \sigma.$$

So the theorem is proved. //

(If $\{A_n\}$ is a sequence of subalgebras of B then $\bigvee_n A_n$ denotes the sub-algebra of B generated by the A_n.)

Corollary 4.8.

If $\{A_n\}$ is an increasing sequence of finite algebras and
$C \overset{\circ}{\subset} \bigvee_n A_n$ then $H(C/A_n) \to 0$ as $n \to \infty$.

Proof: Let $B_0 = \bigcup_{m=1}^{\infty} A_m$; B_0 is an algebra and $C \overset{\circ}{\subset} \sigma(B_0)$ by
hypothesis. By Theorem 4.8 $\forall \ \varepsilon > 0 \ \exists$ finite $D_\varepsilon \subseteq B_0 \ni H(C/D_\varepsilon) < \varepsilon$.
But $D_\varepsilon \subseteq A_{m_0}$ for some m_0 since D_ε is finite, so, if $m \geq m_0$

$$H(C/A_m) \leq H(C/A_{m_0}) \leq H(C/D_\varepsilon) < \varepsilon.$$

Thus, $H(C/A_m)$ tends to zero as $m \to \infty$. //

For entropy to be useful we require methods of calculating its
value. The following is one of the main tools for calculating entropy.

Theorem 4.9: (Kolmogorov-Sinai Theorem)

Let T: (X,B,m) → (X,B,m) be an invertible measure-preserving

transformation and let A be a finite subalgebra of B ∋

$$\bigvee_{n=-\infty}^{\infty} T^n A \overset{\circ}{=} B.$$ Then h(T) = h(T,A).

Proof: Let C ⊆ B be finite. We want to show that

$$h(T,C) \le h(T,A).$$

$$h(T,C) \le h(T, \bigvee_{i=-m}^{m} T^i A) + H(C/ \bigvee_{i=-m}^{m} T^i A)$$

by (4) of Theorem 4.5,

$$= h(T,A) + H(C/ \bigvee_{i=-m}^{m} T^i A)$$

by (5) of Theorem 4.5.

Let $A_m = \bigvee_{i=-m}^{m} T^i A$. It suffices to show that $H(C/A_m)$ goes to zero

as m → ∞. This follows by Corollary 4.8. //

A similar result holds when T is not necessarily invertible:-

Theorem 4.10:

If T: X → X is a measure-preserving transformation (but not

necessarily invertible) and if A is a finite algebra contained in B

with $\bigvee_{i=0}^{\infty} T^i \overset{\circ}{=} B$ then h(T) = h(T,A).

Proof: This is similar to the previous theorem; use $\bigvee_{i=0}^{m-1} T^{-i} A$

in the place of $\bigvee_{i=-m}^{m} T^i A$, and the formula

$$h(T,A) = h(T, \bigvee_{i=0}^{m-1} T^{-i} A)$$

(the proof of which is similar to (5) of Theorem 4.5). //

The following is sometimes useful in showing transformations have zero entropy. We shall use it later to show a rotation of the unit circle has zero entropy.

Corollary 4.10:

If T is invertible and $\bigvee_{i=0}^{\infty} T^{-i}A \stackrel{\circ}{=} B$ for some finite A then $h(T) = 0$.

Proof: By Theorem 4.10

$$h(T) = h(T,A)$$

$$= \lim_{n \to \infty} H(A/\bigvee_{i=1}^{n} T^{-i}A) \quad \text{by Theorem 4.6.}$$

But $\bigvee_{i=1}^{\infty} T^{-i}A \stackrel{\circ}{=} T^{-1}B = B$. Let $A_n = \bigvee_{i=1}^{n} T^{-i}A$; then $A_1 \subset A_2 \subset \dots$

and $\bigvee_{n=1}^{\infty} A_n \stackrel{\circ}{=} B$. By Corollary 4.8 $H(A/A_n) \to 0$ and $h(T) = 0$. //

Remarks:

Entropy can be defined for any countable partition of (X,B,m) as follows: If $\xi = \{A_1, A_2, \dots\}$ then

$$H(\xi) = -\sum_{i} m(A_i)\log m(A_i)$$

(which may be infinite).

A countable partition ξ of X is called a __generator__ for an invertible measure-preserving transformation T if

$$\bigvee_{n=-\infty}^{\infty} T^n A(\xi) \stackrel{\circ}{=} B.$$

As in Theorem 4.9, one can prove that if ξ is a generator and $H(\xi) < \infty$ then $h(T) = h(T,\xi)$.

The basic theorem on existence of generators ξ with $H(\xi) < \infty$ was given by Rohlin in 1963. To state it we need the following definitions: We say that (X,B,m) is __countably__ __generated__ if there exists a countable collection $\{B_n\}$ of elements of B such that the

σ-algebra generated by the $\{B_n\} \stackrel{\circ}{=} B$. (X,B,m) is <u>complete</u> if every subset of a set of measure zero is measurable. Suppose $T: X \to X$ is a measure-preserving transformation, and X is countably generated and complete. We say that T is <u>aperiodic</u> if

$$m(\bigcup_{\substack{n \in Z \\ n \neq 0}} \{x \in X: T^n(x) = x\}) = 0.$$

(The countably generated and completeness conditions ensure that this set is measurable.) Note that T ergodic implies T is aperiodic (unless the space is finite).

<u>Theorem</u> 4.11: (Rohlin [2], 1963)

Suppose (X,B,m) is a countably generated complete non-atomic probability space and $T: X \to X$ is an invertible measure-preserving transformation. Then T has a generator ξ with $H(\xi) < \infty$ iff $h(T) < \infty$ and T is aperiodic.

Thus, if T is ergodic and $h(T) < \infty$ then T has a generator ξ with $H(\xi) < \infty$.

Recently Krieger [1] proved:

<u>Theorem</u> 4.12:

If (X,B,m) is countably generated and T is an invertible ergodic measure-preserving transformation such that $h(T) < \infty$ then T has a finite generator

$$\xi = \{A_1,\ldots,A_n\}.$$

In fact ξ may be taken so that $e^{h(T)} \leq n \leq e^{h(T)} + 1$.

Hence finite generators exist in the most interesting cases although they may be difficult to find.

We now prove some more results that are useful for computation of entropy.

Theorem 4.13:

If B_0 is an algebra and $\sigma(B_0) \cong B$ then

$$h(T) = \sup h(T,A)$$

where the supremum is taken over all finite subalgebras A of B_0.

Proof: Let $\varepsilon > 0$. Let $C \subseteq B$ be finite. By Theorem 4.8 there exists a finite $D_\varepsilon \subseteq B_0$ such that

$$H(C/D_\varepsilon) < \varepsilon.$$

Thus, $\quad h(T,C) \leq h(T,D_\varepsilon) + H(C/D_\varepsilon) \quad$ by (4) of
$$\text{Theorem 4.5}$$
$$\leq h(T,D_\varepsilon) + \varepsilon.$$

Therefore, $\quad h(T,C) \leq \sup_{\substack{D \subseteq B_0 \\ D \text{ finite}}} h(T,D) + \varepsilon,$

and thus $\quad h(T) \leq \sup_{\substack{D \subseteq B_0 \\ D \text{ finite}}} h(T,D).$

The opposite inequality is obvious. //

Theorem 4.14:

Let A_n be finite subalgebras of B such that $A_1 \subseteq A_2 \subseteq \ldots$ and $\bigvee_{n=1}^{\infty} A_n \cong B$. Then $h(T) = \lim_{n \to \infty} h(T,A_n)$.

Proof: We note that $h(T,A_n)$ is an increasing sequence by (3) of Theorem 4.5. $B_0 = \bigcup_{n=1}^{\infty} A_n$ is an algebra and $\sigma(B_0) \cong B$. By Theorem 4.13 $h(T) = \sup_{\substack{C \subseteq B_0 \\ C \text{ finite}}} h(T,C)$. If $C \subseteq B_0$ is finite then $C \subseteq A_{n_0}$ for some n_0. Thus

$$h(T,C) \leq h(T,A_{n_0}).$$

which implies $\qquad h(T) \leq \lim_{n \to \infty} h(T, A_n)$

and hence $\qquad h(T) = \lim_{n \to \infty} h(T, A_n).$ //

Theorem 4.15:

$$h(T_1 \times T_2) = h(T_1) + h(T_2).$$

Proof: Let $T_1 : (X_1, B_1, m_1) \to (X_1, B_1, m_1)$ and

$$T_2 : (X_2, B_2, m_2) \to (X_2, B_2, m_2).$$

If $A_1 \subseteq B_1$, $A_2 \subseteq B_2$ are finite then $A_1 \times A_2$ is finite;

$$\xi(A_1 \times A_2) = \{A_1 \times A_2 : A_1 \in \xi(A_1), \; A_2 \in \xi(A_2)\}.$$

Let

$$T_0 = \bigcup_{\substack{A_1 \subseteq B_1, A_2 \subseteq B_2 \\ \text{finite}}} A_1 \times A_2$$

$$= \text{the algebra of finite unions}$$
$$\text{of measurable rectangles.}$$

Thus, $\sigma(T_0) = B_1 \times B_2$ by definition of $B_1 \times B_2$, and by Theorem 4.13,

$$h(T_1 \times T_2) = \sup_{\substack{C \subseteq T_0 \\ C \text{ finite}}} h(T_1 \times T_2, C).$$

But if C is finite, $C \subseteq T_0$ then $C \subseteq A_1 \times A_2$ for some finite $A_1 \subseteq B_1$, $A_2 \subseteq B_2$. Hence

$$h(T_1 \times T_2) = \sup_{\substack{A_1 \subseteq B_1 \\ A_2 \subseteq B_2 \\ A_1, A_2 \text{ finite}}} h(T_1 \times T_2, A_1 \times A_2).$$

$$H(\bigvee_{i=0}^{n-1} (T_1 \times T_2)^{-i}(A_1 \times A_2))$$

$$= H((\bigvee_{i=0}^{n-1} T_1^{-i} A_1) \times (\bigvee_{i=0}^{n-1} T_2^{-i} A_2))$$

$$= - \sum (m_1 \times m_2)(C_k \times D_j) \cdot \log (m_1 \times m_2)(C_k \times D_j)$$

where the C_k are the members of $\xi(\bigvee_{i=0}^{n-1} T_1^{-i} A_1)$,

and the D_j are the members of $\xi(\bigvee_{i=0}^{n-1} T_2^{-i} A_2)$

$$= - \sum m_1(C_k) m_2(D_j) \cdot \log(m_1(C_k) m_2(D_j))$$

$$= - \sum m_1(C_k) m_2(D_j) \cdot [\log m_1(C_k) + \log m_2(D_j)]$$

$$= - \sum m_1(C_k) \cdot \log m_1(C_k) - \sum m_2(D_j) \cdot \log m_2(D_j)$$

$$= H(\bigvee_{i=0}^{n-1} T_1^{-i} A_1) + H(\bigvee_{i=0}^{n-1} T_2^{-i} A_2).$$

Thus
$$h(T_1 \times T_2, A_1 \times A_2) = h(T_1, A_1) + h(T_2, A_2)$$

so,
$$h(T_1 \times T_2) = h(T_1) + h(T_2). \qquad //$$

§6. Examples

We shall now calculate the entropy of our examples.

(1) If $I: (X, B, m) \to (X, B, m)$ is the identity, then $h(I) = 0$. This is because $h(I, A) = \lim \frac{1}{n} H(A) = 0$. Also, if $T^p = I$ for some $p \neq 0$ then $h(T) = 0$. This follows since $0 = h(T^p) = p \cdot h(T)$ by Theorem 4.7. Hence any measure-preserving transformation of a finite space has zero entropy.

(2) Let $T: K \to K$ be $T(z) = az$.

Case 1: Suppose $\{a^n: n \in Z\}$ is not dense, i.e., a is a root of unity. Thus $a^p = 1$ for some $p \neq 0$; and $T^p(z) = a^p z = z$ so $h(T) = 0$ by example (1).

<u>Case</u> 2: Suppose $\{a^n : n \in Z\}$ is dense in K. Then $\{a^n : n < 0\}$ is dense in K. Let $\xi = \{A_1, A_2\}$ where

A_1 = upper half circle $[1, -1)$

A_2 = lower half circle $[-1, 1)$

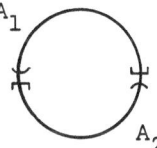

For $n > 0$ $T^{-n}\xi$ consists of semi-circles beginning at a^{-n} and $-a^{-n}$. Since $\{a^{-n} : n > 0\}$ is dense any semi-circle belongs to $\bigvee_{n=0}^{\infty} T^{-n}A(\xi)$. Hence any arc belongs to $\bigvee_{n=0}^{\infty} T^{-n}A(\xi)$. Thus,

$\mathcal{B} = \bigvee_{n=0}^{\infty} T^{-n}A(\xi)$ and so, $h(T) = 0$ by Corollary 4.10. //

(3) Any rotation of a compact metric abelian group has entropy zero.

<u>Proof</u>: (a) Suppose $X = K^n$, the n-torus, and $T(z_1, \ldots, z_n) = (a_1 z_1, \ldots, a_n z_n)$. Then $T = T_1 \times T_2 \times \ldots \times T_n$ where $T_i : K \to K$ is defined by $T_i(z) = a_i z$. By example (2) $h(T_i) = 0$ for all i so by Theorem 4.15

$$h(T) = \sum_{i=1}^{n} h(T_i) = 0.$$

(b) <u>General Case</u>. Let $T: G \to G$ be $T(x) = ax$. Let $\hat{G} = \{\gamma_1, \gamma_2, \ldots\}$. Let $H_n = \text{Ker } \gamma_1 \cap \ldots \cap \text{Ker } \gamma_n$. H_n is a closed subgroup of G and $(\widehat{G/H_n})$ is the group generated by $\{\gamma_1, \ldots, \gamma_n\}$. Thus

$$(\widehat{G/H_n}) = \text{finite group} \times Z^{i_n},$$

so

$$G/H_n = F_n \times K^{i_n}$$

where F_n is a finite group and K^{i_n} is a finite-dimensional torus.

T induces a map $T_n : G/H_n \to G/H_n$ by $T_n(gH_n) = agH_n$. T_n is a rotation on G/H_n, so that it can be written $T_n = T_{n,1} \times T_{n,2}$ where $T_{n,1}$ is a rotation of F_n and $T_{n,2}$ is a rotation of K^{i_n}. Thus,

$$h(T_n) = h(T_{n,1}) + h(T_{n,2}) = 0$$

by example (1) and case (a) of this proof.

Note that $H_n \searrow \{e\}$ so that $\bigvee_n A(G/H_n) \stackrel{\circ}{=} B$ where $A(G/H_n)$ denotes the σ-algebra consisting of those elements of B that are unions of cosets of H_n.

Therefore, if $B_0 = \bigcup_{n=1}^{\infty} A(G/H_n)$ then by Theorem 4.13

$$h(T) = \sup_{\substack{C \subseteq B_0 \\ C \text{ finite}}} h(T,C).$$

However, if $C \subseteq B_0$ is finite then $C \subseteq A(G/H_n)$ for some n and so, $h(T,C) \le h(T_n) = 0$. Thus, $h(T) = 0$. //

Corollary:

Any ergodic transformation with pure point spectrum has zero entropy.

This follows from Theorem 3.4. (Actually we have shown the result only when $L^2(m)$ is separable since the above calculation was for a metric group G.)

(4) Endomorphisms of compact groups:

If A is an endomorphism of the n-torus K^n onto K^n we shall show in Chapter 6 that $h(A) = \sum \log|\lambda|$ where the summation is over all eigenvalues of the matrix $[A]$ with absolute value greater than one.

One can write down a complicated formula for the entropy of an endomorphism of a general compact metric abelian group. See Juzvinskii [1].

(5) Affine transformations:

We shall show in Theorem 6.10 that when $T = a \cdot A$ acts on K^n then $h(T) = h(A)$.

(6) The two-sided $(p_0, \ldots, p_{k-1})$-shift has entropy

$$- \sum_{i=0}^{n-1} p_i \cdot \log p_i.$$

Let $X = \prod_{-\infty}^{\infty} \{0, 1, \ldots, k-1\}$ and T be the shift. Let $A_i = \left\{ \{x_k\} : x_0 = i \right\}$, $0 \leq i \leq k-1$. Then $\xi = \{A_0, \ldots, A_{k-1}\}$ is a partition of X. Let $A = A(\xi)$. By definition of B,

$$\bigvee_{i=-\infty}^{\infty} T^i A = B.$$

By the Kolmogorov-Sinai Theorem (4.9),

$$h(T) = \lim_{n \to \infty} \frac{1}{n} H(A \vee T^{-1}A \vee \ldots \vee T^{-(n-1)}A).$$

A typical element of $\xi(A \vee T^{-1}A \vee \ldots \vee T^{-(n-1)}A)$ is

$$A_{i_0} \cap T^{-1}A_{i_1} \cap \ldots \cap T^{-(n-1)}A_{i_{n-1}}$$

$$= \left\{ \{x_n\} : x_0 = i_0, \ x_1 = i_1, \ldots, x_{n-1} = i_{n-1} \right\}$$

which has measure $p_{i_0} \cdot p_{i_1} \cdots p_{i_{n-1}}$. Thus,

$$H(A \vee T^{-1}A \vee \ldots \vee T^{-(n-1)}A)$$

$$= - \sum (p_{i_0} \cdots p_{i_{n-1}}) \cdot \log(p_{i_0} \cdots p_{i_{n-1}})$$

$$= - \sum_{i_0 = \ldots = i_{n-1} = 0}^{k-1} (p_{i_0} \cdots p_{i_{n-1}})[\log p_{i_0} + \ldots + \log p_{i_{n-1}}]$$

$$= - n \sum_{i=0}^{k-1} p_i \cdot \log p_i.$$

Therefore, $h(T) = h(T,\dot{A}) = - \sum\limits_{i=0}^{k-1} p_i \cdot \log p_i$.   //

Corollary:

The 2-sided $(1/2,1/2)$-shift has entropy log 2; the 2-sided $(1/3,1/3,1/3)$-shift has entropy log 3. Thus these transformations cannot be conjugate.

(7) The 1-sided $(p_0,\ldots,p_{k-1})$-shift has entropy

$$- \sum\limits_{i=0}^{k-1} p_i \cdot \log p_i.$$

The proof is very similar to the one in example (6) but Theorem 4.10 is used instead of Theorem 4.9.

An example of a transformation with infinite entropy is the following:

(8) Let I = (0,1] with Borel sets and Lebesgue measure. Let $X = \prod\limits_{-\infty}^{\infty} I$ with product measure and let T be the shift on X. Then $h(T) = \infty$.

To see this let $A_{n,i} = \left\{ \{x_j\}: \frac{i-1}{n} < x_0 \leq \frac{i}{n} \quad n > 0 \quad 1 \leq i \leq n \right\}$.

Then $m(A_{n,i}) = 1/n$ and $\xi_n = \{A_{n,1},\ldots,A_{n,n}\}$ is a partition of X.

Hence $h(T,\xi_n) = \log n$ by the same argument as used in example (6) (using the independence of $\xi_n, T^{-1}\xi_n,\ldots,T^{-k}\xi_n$). Therefore, $h(T) \geq \log n$ for each n, and so $h(T) = \infty$. //

§7. How good an invariant is entropy?

Definition 4.5:

An invariant P for an equivalence relation is a complete invariant if whenever T and S both have the property P then T and S are equivalent.

Entropy is, in general, far from complete.

(a) An example of two ergodic measure-preserving transformations with equal entropy which are not conjugate.

Let $T: K \to K$ be defined by $T(z) = az$, $a \in K$, where $\{a^n\}$ is dense in K, and let $S: K \to K$ be defined by $S(z) = bz$, $b \in K$, where $\{b^n\}$ is dense in K. T and S are ergodic and $h(T) = 0 = h(S)$ by example (2) of the previous section. If we choose a,b so that $\{a^n\}_{-\infty}^{\infty} \neq \{b^n\}_{-\infty}^{\infty}$ then T and S are not conjugate (in fact, they are not even spectrally equivalent) by Theorem 3.1. //

(b) An example (due to Anzai) of two ergodic and spectrally equivalent measure-preserving transformations with equal entropy which are not conjugate.

Let $T: K^2 \to K^2$ and $S: K^2 \to K^2$ be defined by

$$T(z,w) = (az, z^p w) \qquad S(z,w) = (az, z^q w)$$

where $\{a^n\}_{-\infty}^{\infty}$ is dense in K and p,q are non-zero integers. Observe that T and S are affine transformations, T and S are ergodic, and $h(T) = h(S) = 0$ by example (5) of §6. By considering the characters of K one can easily show that $L^2(m)$ has a basis of the form $\{g_n: n \geq 0\} \cup \{U_T^j f_q: j \in Z, q \geq 0\}$ where $g_n(T) = a^n g_n$. Similarly $L^2(m)$ has a basis $\{g_n\} \cup \{U_S^j h_q: j \in Z, q > 0\}$. One then defines a unitary operator $W: L^2(m) \to L^2(m)$ by $W(g_n) = g_n$ and $W(U_T^j f_q) = U_S^j h_q$ and extending. Clearly $WU_T = U_S W$ showing T and S are spectrally isomorphic.

However if $p \neq \pm q$ T and S are not conjugate. As mentioned before conjugacy and isomorphism coincide for measure-preserving transformations of K equipped with completed Haar measure m. We shall show T and S are not isomorphic. Suppose $\phi T = S\phi$ and $\phi(z,w) = (f(z,w), g(z,w))$. f and g are only defined almost everywhere but this will not affect our argument as we shall consider them as members of $L^2(m)$. We have $f(T) = af$ and $g(T) = f^q g$. Since f is an eigenfunction with eigenvalue a, by remark (3) of §1 of Chapter 3, $f(z,w) = c \cdot z$ for some $c \in K$. The second equation then becomes $g(T(z,w)) = c^q z^q g(z,w)$. If one now expresses g as a Fourier series, then it is straightforward to show that $g(z,w) = kz^n w^m$ where $k \in K$, $pm = q$ for some $m \in Z$ and $a^n = c^q$. So $\phi(z,w) = (cz, kz^n w^m)$ is an affine transformation and for ϕ to be an invertible measure-preserving transformation one needs $m = \pm 1$, i.e., $p = \pm q$. //

However we can consider the problem of completeness of entropy for certain collections of measure-preserving transformations, and this we do in the next section.

§8. <u>Bernoulli</u> <u>and</u> <u>Kolmogorov</u> <u>Automorphisms</u>

(As general references for this section see Shields [2] and Friedman and Ornstein [2].)

<u>Definition</u> 4.6:

Let (Y,F,μ) be a probability space. Let

$$(X,B,m) = \prod_{-\infty}^{\infty} (Y,F,\mu)$$

and let $T: X \rightarrow X$ be the shift

$$T(\{y_n\}) = \{x_n\} \quad \text{where} \quad x_n = y_{n+1} \quad n \in Z.$$

T is an invertible measure-preserving transformation and is called the <u>Bernoulli</u> automorphism with <u>state</u> <u>space</u> (Y,F,μ).

<u>Examples</u> <u>of</u> Bernoulli automorphisms:

(1) the 2-sided $(p_0, \ldots, p_{k-1})$-shift. Here $Y = \{0,1,\ldots,k-1\}$.

(2) the example (8) of §6. Here $y = (0,1]$.

(3) If T is a Bernoulli automorphism so is T^2.

(4) If T_1 and T_2 are Bernoulli automorphisms so is $T_1{\times}T_2$.

Remark:

If T is a Bernoulli automorphism then $h(T) < \infty$ iff ∃ a countable partition η on (Y,F,μ) ∍ $H(\eta) < \infty$ and $A(\eta) = F$. In this case $h(T) = H(F)$.

We shall call a probability space a Lebesgue space if the identity map of it is isomorphic (as a measure-preserving transformation) to the identity map on a probability space consisting of a subinterval of [0,1] with Lebesgue measurable sets and Lebesgue measure together with some atoms.

Recently entropy has been shown to be a complete invariant for the class of Bernoulli automorphisms:

Theorem 4.18: (Ornstein [1] and [2])

Two Bernoulli automorphisms (whose state spaces are Lebesgue spaces) with the same entropy are conjugate; i.e., entropy is a complete invariant for the conjugacy of Bernoulli automorphisms.

(Isomorphism follows from conjugacy by the assumption on the state spaces.)

Certain special cases had been worked out earlier by Meshalkin [1] and by Blum and Hanson [1]. This result reduces the conjugacy problem for Bernoulli automorphisms to their state spaces, since the entropy depends only on the state space. It is possible, for example, for a Bernoulli automorphism with a state space of two points to be conjugate to a Bernoulli automorphism with a countably infinite state space.

Note:

Given any $x > 0$ one can find $n > 0$ and $\{p_1,p_2,\ldots,p_n\}$, $p_i \geq 0$ $\sum_{i=0}^{n} p_i = 1$ such that $- \sum_{i=0}^{n} p_i \cdot \log p_i = x$. Hence ∀ $x > 0$

∃ a Bernoulli automorphism with entropy x.

Corollary 4.16:

(i)   Every Bernoulli automorphism has an n-th root.

(ii) Every Bernoulli automorphism can be written as a direct product of two Bernoulli automorphisms.

Proof: (i) Let T be a Bernoulli automorphism and n > 0. Let S be a Bernoulli automorphism with $h(S) = \frac{1}{n} h(T)$. Then S^n is a Bernoulli automorphism with entropy h(T), and therefore S^n and T are conjugate.

(ii) Let T be a Bernoulli automorphism. Let S be Bernoulli with h(S) = ½·h(T). Then h(S×S) = h(T) and since S×S is Bernoulli, S×S is conjugate to T. //

Ornstein has proved many deep results about Bernoulli automorphisms, some of which we now summarize:

Theorem 4.17: (Ornstein [3] etc.)

(i)   Every root of a Bernoulli automorphism is a Bernoulli automorphism. (S is an n-th root if $S^n = T$.)

(ii) Let T be a Bernoulli automorphism. If F is a sub-σ-algebra of B with TF = F then T restricted to the measure space $(X,F,m|_F)$ is a Bernoulli automorphism (i.e., a factor of a Bernoulli automorphism is a Bernoulli automorphism).

(iii) If $F_n \nearrow B$ is an increasing sequence of σ-algebras with $TF_n = F_n$ and T restricted to $(X,F_n,m|_{F_n})$ is a Bernoulli automorphism ∀ n > 0, then T is a Bernoulli automorphism (i.e., an inverse limit of Bernoulli automorphisms is a Bernoulli automorphism).

The following class of transformations were introduced by Kolmogorov in 1958 by analogy with regular stochastic processes.

Definition 4.7:

An invertible measure-preserving transformation T of a probability space (X, B, m) is a <u>Kolmogorov automorphism</u> (K-automorphism) if $\exists$ a sub-σ-algebra K of B such that:

(i) $\qquad\qquad\qquad\qquad K \overset{\circ}{\subset} TK.$

(ii) $\qquad\qquad\qquad\qquad \overset{\infty}{\underset{n=0}{\bigvee}} T^n K \overset{\circ}{=} B.$

(iii) $\qquad\qquad\qquad\qquad \overset{\infty}{\underset{n=0}{\bigcap}} T^{-n} K \overset{\circ}{=} N = \{X, \phi\}.$

(If A, C are σ-algebras $A \overset{\circ}{\subset} C$ will mean $\forall A \in A$ $\exists$ $C \in C$ with $m(A \triangle C) = 0$. If $B_1, B_2 \in B$ then $B_1 \overset{\circ}{\subset} B_2$ means $m(B_1 \setminus B_2) = 0$ and $B_1 \overset{\circ}{=} B_2$ means $m(B_1 \triangle B_2) = 0$.)

Theorem 4.18:

Every Bernoulli automorphism is a Kolmogorov automorphism.

<u>Proof</u>: Let the state space for T be (Y, F, μ). If $F \in F$, let $\tilde{F} = \left\{ \{x_n\} \in X : x_0 \in F \right\} \in B$. Let $G = \{\tilde{F} : F \in F\}$, which is called the <u>time-0 σ-algebra</u>. Let $K = \overset{0}{\underset{i=-\infty}{\bigvee}} T^i G$. We now verify that K satisfies the conditions for a Kolmogorov automorphism.

(i) $\qquad\qquad K = \overset{0}{\underset{i=-\infty}{\bigvee}} T^i G \subset \overset{1}{\underset{i=-\infty}{\bigvee}} T^i G = TK.$

(ii) $\overset{\infty}{\underset{n=0}{\bigvee}} T^n K = \overset{\infty}{\underset{n=0}{\bigvee}} \overset{n}{\underset{i=-\infty}{\bigvee}} T^i G = \overset{\infty}{\underset{-\infty}{\bigvee}} T^i G = B$ by definition of B.

(iii) We have to show $\overset{\infty}{\underset{0}{\bigcap}} T^{-n} K \overset{\circ}{=} N = \{X, \phi\}$. Fix $A \in \overset{\infty}{\underset{0}{\bigcap}} T^{-n} K = \overset{\infty}{\underset{n=0}{\bigcap}} \overset{-n}{\underset{-\infty}{\bigvee}} T^i G$. Let $B \in \overset{\infty}{\underset{k=j}{\bigvee}} T^k G$, $j \in Z$. Since $A \in \underset{i<j}{\bigvee} T^i G$, A and B are independent, and therefore $m(A \cap B) = m(A) m(B)$. The collection of all sets B for which $m(A \cap B) = m(A) m(B)$ is a monotone class,

and, by the above, contains $\bigcup_{j=-\infty}^{\infty} \bigvee_{k=j}^{\infty} T^k G$. Therefore $\forall\ B \in \mathcal{B}$,

$m(A \cap B) = m(A)m(B)$. Put $B = A$, then $m(A) = m(A)^2$ which implies

$m(A) = 0$ or 1. Hence

$$\bigcap_{n=0}^{\infty} T^{-n} K \overset{o}{=} N. \quad //$$

It was an open problem from 1958 to 1969 as to whether the converse of Theorem 4.18 was true, i.e., whether a Kolmogorov automorphism is conjugate to a Bernoulli automorphism. This was shown to be false by Ornstein.

Theorem 4.19: (Ornstein [6])

There is an example of a Kolmogorov automorphism T with the following properties:

(i) T is not conjugate to a Bernoulli automorphism.

(ii) T does not have a square root.

(iii) T is not conjugate to T^{-1}.

Corollary 4.19:

Entropy is not a complete invariant for the class of Kolmogorov automorphisms.

Proof: Let T be the example of Ornstein. By Theorem 4.9 $h(T) > 0$. Choose a Bernoulli automorphism S with $h(S) = h(T)$. S and T are not isomorphic by (i).

Remarks:

(1) Property (iii) of T contrasts with the behavior of ergodic transformations with pure point spectrum. (See Corollary 3.3.)

(2) Ornstein's example is defined by induction and so is fairly complicated to describe. It is therefore important to check whether the more "natural" examples of Kolmogorov automorphisms are Bernoulli

automorphisms or not. We consider some of these at the end of this
section.

(3) Sinai has proved that if T is an ergodic invertible measure-
preserving transformation of (X,B,m), which we assume is isomorphic
to the unit interval with Lebesgue measure, and h(T) > 0 and if S
is a Bernoulli automorphism with h(S) ≤ h(T) then there exists a
measure-preserving transformation φ such that φT = Sφ, i.e., S
is a factor of T.

The next theorem shows that Kolmogorov automorphisms and Bernoulli automorphisms are spectrally the same:

Theorem 4.20: (Rohlin)

A Kolmogorov automorphism T of a probability space (X,B,m),
where $L^2(m)$ is separable, has countable Lebesgue spectrum.

Proof: Assume B ≠ {X,φ} = N or else the result is trivial.
We have (i) $K \overset{\circ}{\subset} TK$ (ii) $\bigvee T^n K \overset{\circ}{=} B$ (iii) $\bigcap T^{-n} K \overset{\circ}{=} N$. We split
the proof into three parts:

a) We first show that K has no atoms, i.e., if C ∈ K and m(C) > 0
then ∃ D ∈ K with D ⊂ C and m(D) < m(C).

Suppose C is an atom of K with m(C) > 0. Then TC is an
atom of TK and since $K \overset{\circ}{\subset} TK$ either $TC \overset{\circ}{\subset} C$ or m(C ∩ TC) = 0. If
$TC \overset{\circ}{\subset} C$ then $TC \overset{\circ}{=} C$ since both sets have the same measure so that
$C \overset{\circ}{\in} \bigcap T^{-n} K$ and therefore m(C) = 1. Hence $B \overset{\circ}{=} \{X,φ\}$, a contradiction. On the other hand, suppose m(TC ∩ C) = 0. Then either for some
k > 0 $T^k C \overset{\circ}{\subset} C$ (and we use the above proof to get a contradiction)
or $m(T^k C ∩ C) = 0$ ∀ k > 0 and then $C ∪ TC ∪ T^2 C ∪ \ldots$ has infinite measure, a contradiction.

b) Let $H = \{f ∈ L^2(m): f \text{ is } K\text{-measurable}\}$. Then $U_T H \subset H$. Let
$H = V \oplus U_T H$. From $U_T^{-n} H = \bigoplus_{-n}^{m} U_T^i V \oplus U_T^{m+1} H$ (n,m > 0) we conclude

that $L^2(m) = \bigoplus_{-\infty}^{\infty} U_T^n V \oplus C$ where C is the subspace of constants. It

suffices to show V is infinite-dimensional since if $\{f_1, f_2, f_3, \ldots\}$

is a basis for V, then $\{f_0 \equiv 1, U_T^n f_m: n \in Z, m > 0\}$ is a basis for

$L^2(m)$.

c) We now show V is infinite-dimensional. Since $TK \overset{\circ}{\neq} K$ (we are

assuming $B \overset{\circ}{\neq} N$) we know $V \neq \{0\}$. Let $g \in V$, $g \neq 0$ and then

$G = \{x: g(x) \neq 0\}$ satisfies $m(G) > 0$. Since $G \in K$ and using a)

we know $\chi_G H = \{\chi_G f: f \in H\}$ is infinite-dimensional. $\chi_G H = V' \oplus \chi_G U_T H$

where $V' \subset V$ so either V' is infinite-dimensional (and hence V

is) or $\chi_G U_T H$ is infinite-dimensional. In this case there is a

linearly independent sequence of functions $\{\chi_G U_T f_n\}$ where the f_n

are bounded functions in H. Then $\{g U_T f_n\}$ are linearly independent

in H. It suffices to show these functions are in V. But if $f \in H$

then

$$(g U_T f_n, U_T f) = (g, U_T(f \bar{f}_n)) = 0$$

so $g U_T f_n \in V$. //

Corollary 4.20:

A K-automorphism is strong mixing.

Proof: By Theorem 2.3. //

Kolmogorov automorphisms are connected to entropy by the following result (half of which was proved by Pinsker).

Theorem 4.21: (Rohlin & Sinai [1])

Let (X, B, m) be isomorphic to $[0,1]$ with Lebesgue measure.
Let $T: X \to X$ be invertible and measure-preserving. Then T is a
K-automorphism iff $h(T,A) > 0$ $\forall$ finite $A \overset{\circ}{\neq} N$; i.e., T has
completely positive entropy.

Remark:

This shows that K-automorphisms are "the opposites" of transformations with zero entropy (since $h(T,A) = 0 \quad \forall \quad A$ in the zero
entropy case).

Examples:

(1) Group Automorphisms. Rohlin proved that any ergodic automorphism
of a compact abelian metric group is a K-automorphism and later
Yusinskii proved the theorem in the non-abelian case. Katznelson [1]
has shown that ergodic automorphisms of finite-dimensional tori are
conjugate to Bernoulli automorphisms. Chu [1] and Lind [1] have independently extended Katznelson's results to the (countably) infinite-
dimensional torus. Katznelson and Weiss [2] have also solved the
case where the dual group is the discrete rationals but whether an
ergodic automorphism of a general compact abelian metric group is
Bernoulli is not yet known.

(2) Markov Chains. Consider a two-sided Markov chain with transition
matrix $[p_{ij}]$. The shift T on the space of sequences of integers
becomes a measure-preserving transformation for the Markov measure defined by $[p_{ij}]$ and an initial vector $[p_1,\ldots,p_k]$ satisfying
$[p_1,\ldots,p_k][p_{ij}] = [p_1,\ldots,p_k]$. (See Billingsley [1].) It is known
that T is ergodic iff the chain is irreducible (i.e., $\forall$ pairs of
states i,j $\exists$ n > 0 with $p_{ij}^{(n)} > 0$) and T is strong mixing iff
the chain is irreducible and aperiodic (i.e., $\forall$ states i
g.c.d. $\{n: p_{ii}^{(n)} > 0\} = 1$). Friedman and Ornstein.[1] have shown
that if T is strong mixing then it is conjugate to a Bernoulli automorphism. Therefore, from the point of view of ergodic theory mixing
Markov chains are the same as Bernoulli automorphisms, i.e., we can
represent the space as a direct product measure space so that T
becomes the shift on the new space.

(3) One can generalize the notion of a finite-dimensional torus to obtain another kind of homogeneous space:--namely a nilmanifold. Let N be a connected, simple connected, nilpotent Lie group and D a discrete subgroup of N so that the quotient space N/D is compact. N/D is called a <u>nilmanifold</u>. When $N = R^n$ and $D = Z^n$ we get an n-torus. The Haar measure on N determines a normalized Borel measure on N/D. If $\bar{A}: N \rightarrow N$ is a (continuous) automorphism with $\bar{A}D = D$ then this induces a map $A: N/D \rightarrow N/D$, which we call an <u>automorphism</u> <u>of</u> <u>N/D</u>. A always preserves the measure m. Parry has investigated the ergodic theory of such maps and has shown that if A is ergodic then A is a K-automorphism. A subclass of the automor-phisms of N/D, the Anosov ones, are known to be conjugate to Ber-noulli automorphisms, but it has not yet been proved that the others are.

The simplest examples are as follows: Let

$$N = \left\{ \begin{pmatrix} 1 & x & z \\ 0 & 1 & y \\ 0 & 0 & 1 \end{pmatrix} : x,y,z \in R \right\}.$$

N satisfies the above conditions with the operation of matrix multi-plication and the natural topology from R^3. Let

$$D = \left\{ \begin{pmatrix} 1 & m & p \\ 0 & 1 & n \\ 0 & 0 & 1 \end{pmatrix} : m,n,p \in Z \right\}.$$

Then N/D is a nilmanifold. The automorphism

$$\begin{pmatrix} 1 & x & z \\ 0 & 1 & y \\ 0 & 0 & 1 \end{pmatrix} \rightarrow \begin{pmatrix} 1 & 2x+y & z+x^2+xy+\frac{y^2}{2} \\ 0 & 1 & x+y \\ 0 & 0 & 1 \end{pmatrix}$$

of N induces an ergodic automorphism of N/D, and it is unknown if this is Bernoulli.

§9. Pinsker Algebra

Let T be a measure-preserving transformation of a probability
space (X,B,m) which is isomorphic to $[0,1]$ with Lebesgue measure.
Let

$$P(T) = \bigvee \{A: \ A \subset B, \ A \text{ finite}, \ h(T,A) = 0\}.$$

This is called the Pinsker σ-algebra.

One can show that $T^{-1}P(T) = P(T)$. One can also prove that if A
is finite then $A \subset P(T)$ iff $h(T,A) = 0$. Thus, $P(T)$ is the maxi-
mum σ-algebra such that T restricted to $(X,P(T),m|_{P(T)})$ has zero
entropy. Note that $P(T) = B$ iff $h(T) = 0$ and $P(T) = N$ iff T
is a K-automorphism (by Theorem 4.21).

Theorem 4.22: (Rohlin)

If T is an invertible measure-preserving transformation with
$h(T) > 0$ then U_T has countable Lebesgue spectrum in

$$L^2(B) \ominus L^2(P(T)).$$

This reduces the study of the spectrum of invertible measure-
preserving transformations to those with zero entropy.

Corollary 4.22:

Transformations with pure point spectrum have zero entropy.

Proof: $L^2(m)$ cannot have a subspace on which U_T has countable
Lebesgue spectrum.

The types of spectrum that can occur for zero entropy transformations are unknown. There are examples of zero entropy transformations with countable Lebesgue spectrum (from Gaussian processes and
horocycle flows).

In the space of invertible measure-preserving transformations of
(X,B,m) with the weak topology, the set of transformations of zero

entropy forms a dense G_δ.

Pinsker [1] conjectured that any ergodic measure-preserving transformation could be written as a direct product of one with zero entropy and a K-automorphism. However, (ii) of Theorem 4.19 allows us to obtain a counterexample for if $T: X \to X$ is the example of Ornstein with no square root then the transformation S of the space $\{0\} \times X \cup \{1\} \times X$ given by $S(0,x) = (1,x)$. $S(1,x) = (0,Tx)$ provides a counterexample to the Pinsker conjecture. This example is not mixing (since S^2 is not ergodic), but Ornstein has constructed a mixing transformation which violates the Pinsker conjecture.

§10. Sequence Entropy
(See: Kushnirenko [1].)

Let (X, B, m) be a probability space isomorphic to $[0,1]$ with Lebesgue measure. Let $T: X \to X$ be an invertible measure-preserving transformation.

Let $\Gamma = \{t_1, t_2, \ldots\}$ be a sequence of integers. Let A be a finite algebra $A \subset B$.

Define $\qquad h_\Gamma(T,A) = \lim_{n \to \infty} \sup \frac{1}{n} H(T^{t_1}A \vee \ldots \vee T^{t_n}A)$

and define $\qquad\qquad h_\Gamma(T) = \sup_{A \text{ finite}} h_\Gamma(T,A).$

It is easily shown that $h_\Gamma(T)$ is a conjugacy invariant for each Γ. Entropy and spectral properties are connected by the following:

Theorem 4.23: (Kushnirenko [1])

T has discrete spectrum iff $h_\Gamma(T) = 0$ $\forall$ Γ.

One can also show that $\sup_\Gamma h_\Gamma(T) = \infty$ or $\log k$, for some $k > 0$, $k \in Z$; moreover, those T with $\sup_\Gamma h_\Gamma(T) = \log k$ are easy to describe.

Problem:

If T has quasi-discrete spectrum (see Hahn and Parry [1]), what
sequences give $h_\Gamma(T) > 0$?

$h_\Gamma(T)$ has been calculated except in the cases when T has zero
entropy and Γ has "large gaps". $h_\Gamma(T)$ will only give new information when $h(T) = 0$. (See: Newton [1].)

§11. Comments

Entropy was introduced as a conjugacy invariant for measure-
preserving transformations. It was soon realized that entropy was
more than just an assignment of a number to each transformation.

Kolmogorov automorphisms and transformations with zero entropy
have received the most treatment. They are "opposites" from the point
of view of entropy. Kolmogorov automorphisms are important for applications as it seems that the most interesting smooth systems are
Kolmogorov and even Bernoulli.

By Theorem 4.22 we know that the spectral theory of invertible
measure-preserving transformations reduces to that for the zero entropy case. The following is still an open problem:

If $h(T) = 0$ what kind of spectrum can U_T have?

For transformations with zero entropy the isomorphism problem is
only solved for ergodic transformations with discrete spectrum, to-
tally ergodic transformations with quasi-discrete spectrum and some
other special cases. Sequence entropy may play a role in the isomorphism problem for zero entropy transformations.

We note again that in the weak topology on the set of all invertible measure-preserving transformations on a given space $(X,\mathcal{B},m)$,
the set of transformations with zero entropy is a dense G_δ (countable intersection of open sets) and the set of weak mixing transformations is also a dense G_δ. Since the set of strong mixing

transformations is a set of first category it follows that "most" transformations are weak mixing, have zero entropy, but are not strong mixing.

The main problem to consider for Kolmogorov automorphisms seems to be to find more examples of Kolmogorov automorphisms that are not conjugate Bernoulli automorphisms. One should first check this fact for all the usual ways of constructing new transformations from old ones (e.g., is a weak mixing group extension of a Bernoulli automorphism a Bernoulli automorphism?). Then one might hope to find a new invariant that may be complete for Kolmogorov automorphisms.

§12. Non-invertible Transformations

Suppose $T: (X,B,m) \to (X,B,m)$ is measure-preserving but not necessarily invertible; assume that (X,B,m) is isomorphic to $[0,1]$ with Lebesgue measure. Note that

$$B \supset T^{-1}B \supset T^{-2}B \supset \ldots \quad .$$

Let $B_\infty = \bigcap_{n=0}^{\infty} T^{-n}B$; so, $T^{-1}B_\infty = B_\infty$, and $T|_{(X,B_\infty,m)}$ is invertible. One can show that U_T has countable Lebesgue spectrum in $L^2(B) \ominus L^2(B_\infty)$ where countable Lebesgue spectrum in this situation means there is a basis of the form

$$\{U^n f_m: n \geq 0 \text{ and } m > 0\}.$$

This reduces the study of spectral properties of measure-preserving transformations to those which are invertible (in fact, invertible ones with zero entropy by Theorem 4.22).

One can also show that $P(T) \overset{\circ}{\subset} B$ (i.e., if $h(T) = 0$ then T is invertible modulo sets of measure zero; more precisely $\underset{\sim}{T}^{-1}$ is a measure algebra isomorphism).

The analogous concept to a K-automorphism is an exact endomorphism.

<u>Definition</u> 4.8:

T: $X \to X$ is an <u>exact</u> <u>endomorphism</u> if

$$\bigcap_{n=0}^{\infty} T^{-n}B \cong N \; ; \quad \text{i.e.,} \quad B_{\infty} \cong N.$$

So exact endomorphisms are as far from being invertible as possible. Examples of exact endomorphisms are the one-sided Bernoulli shifts. Exact endomorphisms are strong mixing (by the above remarks about spectrum and a proof like that of Theorem 2.3).

It was conjectured that every ergodic measure-preserving transformation is a product of an exact endomorphism and an invertible measure-preserving transformation. This is not so (Parry and Walters).

Also, one-sided Bernoulli shifts with the same entropy are not necessarily conjugate since an m-to-1 map cannot be conjugate to an n-to-1 map if $m \neq n$. So entropy is far from complete for exact endomorphisms. Parry and Walters (1971) constructed two exact endomorphisms S,T with $S^{-n}B = T^{-n}B \; \forall \; n > 0$, $S^2 = T^2$ ($\Rightarrow$ h(S) = h(T)) but with S and T not conjugate. The method used involved the Jacobian of an endomorphism, a concept which was introduced in Parry [3]. (It is not known if there are two K-automorphisms S,T with $S^2 = T^2$ but with S and T not conjugate.) Also, exact endomorphisms need not be conjugate to one-sided Bernoulli shifts; in fact a one-sided Markov chain which is exact need not be conjugate to a one-sided Bernoulli shift.

Chapter 5: Topological Dynamics

§0. Introduction

In measure theoretic ergodic theory one studies the asymptotic properties of measure-preserving transformations. In topological dynamics one studies the asymptotic properties of continuous maps.

Theorem 5.0:

Let X be a compact Hausdorff space. The following are equivalent:

(1) X is metrizable.

(2) X has a countable base.

(3) C(X) (the space of all complex-valued continuous functions on X) has a countable dense subset.

Proof: See Kelley [1].

We shall consider compact metric spaces X and homeomorphisms T: X → X. C(X) is a Banach algebra with

$$\|f\| = \sup_{x \in X} |f(x)|.$$

The map U_T: C(X) → C(X), defined by $(U_T f)(x) = f(Tx)$ is clearly a multiplicative linear isometry of C(X) onto C(X), i.e., U_T is a Banach algebra automorphism.

Remarks:

Compactness is a "finiteness" condition and corresponds to the assumption of a finite measure in the measure theoretic work. The assumption of metrizability is not needed for many of the results but it often shortens proofs and most applications are for metric spaces.

We assume that T is a homeomorphism, rather than a continuous map, for simplicity.

Typical examples that we shall study are:

Examples:

(i) I on any X.

(ii) $Tx = ax$ where X is a compact metric group. (On such a group there exists a left invariant metric d, i.e.,

$$d(bx,by) = d(x,y) \quad \forall \quad b,x,y \in X \quad).$$

(iii) an automorphism of a compact metric group.

(iv)   an affine transformation $Tx = a \cdot A(x)$ where A is an automorphism of a compact group G and $a \in G$.

(v)   Let $Y = \{0,1,\ldots,k-1\}$ with the discrete topology. Let $X = \prod_{-\infty}^{\infty} Y$ with the product topology. A metric on X is given by:

$d(\{x_n\},\{y_n\}) = \sum_{n=-\infty}^{\infty} \frac{|x_n-y_n|}{2^{|n|}}$. The shift T ($T\{x_n\} = \{y_n\}$ with $y_n = x_{n+1}$) is a homeomorphism of X. Note that here we have a special case of (iii) since X is a compact group under the operation

$$\{x_n\} + \{y_n\} = \{(x_n+y_n)\bmod(k)\},$$

and T is an automorphism of X. d is an invariant metric on X.

§1. Minimality

X will denote a compact metric space and $T: X \to X$ a homeomorphism. We would like to find a concept of "irreducible piece" to play the role ergodicity played for measure-preserving transformations.

Definition 5.1:

T is <u>minimal</u> if $\forall$ $x \in X$ $\{T^n x: n \in Z\}$ is dense in X. $O_T(x) = \{T^n x: n \in Z\}$ is called the <u>T-orbit of</u> <u>x</u>.

Theorem 5.1:

   T is minimal iff TE = E and E closed ⇒ E = φ or X.

   Proof: Suppose T is minimal and suppose E is closed, E ≠ φ
and TE = E. Choose x ∈ E. Then $O_T(x)$ ⊂ E by the T-invariance
of E, and X = $\overline{O_T(x)}$ so $\overline{E}$ = X i.e., E = X. Conversely, ∀
x ∈ X, $\overline{O_T(x)}$ is a closed T-invariant non-empty set, and hence is
all of X. //

Definition 5.2:

   A closed subset E of X which is T-invariant is called a
minimal set with respect to T: X → X if $T|_E$ is minimal.

Theorem 5.2:

   Any homeomorphism T: X → X has a minimal set.

   Proof: Let E denote the collection of all closed non-empty
T-invariant subsets of X. Clearly E ≠ φ since X belongs to E.
E is a partially ordered set under inclusion. Every linearly ordered
subset of E has a least element (the intersection of the elements of
the chain. The least element is non-empty by Cantor's intersection
property.) Thus, by Zorn's Lemma, E has a minimum element. This
element is a minimal set for T. //

Remark:

   Ergodicity has the properties:
(i)   An ergodic transformation is indecomposable.
(ii) Every measure-preserving transformation on a decent space can be
decomposed into ergodic pieces in a nice way.
   By its definition, a minimal transformation is indecomposable.
We know that each homeomorphism T: X → X has a minimal set. However,
in general, one cannot partition X into T-invariant closed sets E_α
such that X = $\bigcup_\alpha E_\alpha$, $TE_\alpha = E_\alpha$ ∀ α and $T|_{E_\alpha}$ is minimal (although

we can in some important cases). If T has such a decomposition it
is sometimes called semi-simple. An example of a transformation not
admitting such a decomposition is an ergodic automorphism of a compact
metric group. This will be clear from the next section.

Definition 5.3:

A point $x \in X$ is a periodic point of T if $T^n x = x$ for some
$n \neq 0$. The least such positive n with this property is called the
period of x under T.

Theorem 5.3:

Let $T: X \to X$ be a minimal homeomorphism. Then:

(1) T has no nonconstant invariant continuous functions.

(2) If X is not finite T has no periodic points.

Proof: (1) $fT(x) = f(x)$ implies $fT^n(x) = f(x)$ $\forall$ $n \in Z$, and
so f is constant on the dense subset $O_T(x)$ of X. Thus f is
constant on X.

(2) If $T^n x = x$, $n \neq 0$ then $\{x, Tx, \ldots, T^{n-1} x\}$ is a closed
T-invariant set and by the minimality condition it is the whole
space X. //

Remarks:

(i)   If T has no nonconstant T-invariant functions then T need
not be minimal. As an example of this, let A be an ergodic automorphism of a compact metric group $G \neq \{e\}$. Then A is not minimal
since $A(e) = e$. But A satisfies property (1), since if $fA(x) = f(x)$, f continuous, then by ergodicity, f = constant a.e. and,
since Haar measure is positive on open sets f is constant every-
where.

(ii) The fact that a minimal homeomorphism of a non-finite space has
no periodic points rules out many important examples, such as ergodic
automorphisms of compact metric groups.

We now check whether the examples mentioned in §0 are minimal or not.

Examples:

(i)   I is minimal iff X = a single point.

(ii)   Let G be a compact metric group and $T(x) = ax$. T is minimal iff $\{a^n : n \in Z\}$ is dense in X.

    Proof: ($\Rightarrow$) $O_T(e) = \{a^n : n \in Z\}$.

    ($\Leftarrow$) Let $x \in X$. We must show that $\overline{O_T(x)} = X$. Let $y \in X$. $\exists \ n_i \ni a^{n_i} \to yx^{-1}$ i.e.,

$$a^{n_i} \cdot x \to y$$
$$\|$$
$$T^{n_i}(x) \to y.$$

Therefore $O_T(x)$ is dense in X. //

(iii) An automorphism of a compact metric group G is minimal iff G = {e}. The proof is trivial.

(iv)   For affine transformations of compact metric groups necessary and sufficient conditions for minimality are known. For example, if G is also abelian and connected then $T = a \cdot A$ is minimal iff

$$\bigcap_{n=0}^{\infty} B^n G = \{e\} \quad \text{and} \quad [a, BG] = G$$

where B is the endomorphism of G defined by $B(x) = x^{-1} \cdot A(x)$ and [a,BG] denotes the smallest closed subgroup of G containing a and BG. This was proved by Hoare and Parry [1].

(v)   The shift on k symbols is minimal iff k = 0. This is seen from (iii) above.

§2. Topological Transitivity

Definition 5.4:

T: $X \to X$ is topologically transitive if $\exists$ $x_0 \in X$ $\ni$ $O_T(x_0)$ is dense in X.

Note:

T minimal $\Rightarrow$ T topologically transitive.

Theorem 5.4:

The following are equivalent:

(1) T is topologically transitive.

(2) TE = E, E closed, $E \neq X \Rightarrow E$ is nowhere dense (or, equivalent-ly, if TU = U, U open, $U \neq \phi$ then U is dense).

(3) If U,V are nonempty open sets then $\exists$ $n \in Z$ $\ni$

$$T^n(U) \cap V \neq \phi.$$

(4) $\{x \in X: \overline{O_T(x)} \neq X\}$ is a set of first category.

Proof: (1) $\Rightarrow$ (2). Suppose $\overline{O_T(x_0)} = X$ and let $E \neq \phi$, E closed, TE = E, $E \neq X$. Suppose $U \subseteq E$ is open, $U \neq \phi$. Then $\exists$ p $\ni$ $T^p(x_0) \in U \subseteq E$ so that $O_T(x_0) \subseteq E$ and X = E, a contra-diction. Therefore E has no interior.

(2) $\Rightarrow$ (3). Suppose $U,V \neq \phi$ are open sets. Then $\bigcup\limits_{n=-\infty}^{\infty} T^n U$ is an open T-invariant set; so, it is necessarily dense by condition (2). Thus $\bigcup\limits_{n=-\infty}^{\infty} T^n U \cap V \neq \phi$.

(3) $\Rightarrow$ (4). Let $U_1, U_2, \ldots, U_n, \ldots$ be a countable base for X. Then $\overline{O_T(x)} \neq X$

$$\Leftrightarrow \exists \ n \ni O_T(x) \cap U_n = \phi$$

$$\Leftrightarrow \exists \ n \ni T^m(x) \in X \backslash U_n \ \forall \ m \in Z$$

$$\Leftrightarrow \exists \ n \ \text{with} \ x \in \bigcap_{n=-\infty}^{\infty} T^m(X \backslash U_n)$$

$$\Leftrightarrow x \in \bigcup_{n=1}^{\infty} \bigcap_{m=-\infty}^{\infty} T^m(X \backslash U_n).$$

It suffices to show $\bigcap_{m=-\infty}^{\infty} T^m(X \backslash U_n)$ is a nowhere dense set $\forall$ n. Its complement is $\bigcup_{m=-\infty}^{\infty} T^m(U_n)$ which is clearly dense by condition (3). Hence the result follows.

(4) $\Rightarrow$ (1). This is clear since a compact metric space is of second category. //

The following theorem gives many examples of topologically transitive homeomorphisms.

Theorem 5.5:

Let X be a compact metric space and T: X → X a homeomorphism, m a Borel probability measure on X giving positive measure to every non-empty open set. If T is ergodic with respect to m, then $m\{x \in X: \overline{O_T(x)} = X\} = 1$. In particular, T is topologically transitive.

Proof: Let $U_1, U_2, \ldots$ be a countable base for the topology. By the previous proof

$$\{x: \overline{O_T(x)} \neq X\} = \bigcup_{n=1}^{\infty} \bigcap_{k=-\infty}^{\infty} T^k(X \backslash U_n).$$

The closed set $\bigcap_{k=-\infty}^{\infty} T^k(X \backslash U_n)$ is T-invariant, so by ergodicity has measure 0 or 1. But U_n is contained in complement of this set and $m(U_n) > 0$, since U_n is open. Therefore

$$m(\bigcap_{k=-\infty}^{\infty} T^k(X \backslash U_n)) = 0$$

and so $m\{x: \overline{O_T(x)} \neq X\} = 0$. Hence $m\{x: \overline{O_T(x)} = X\} = 1$. //

Corollary 5.5:

Let G be a compact metric group and T: G → G an affine transformation. T is ergodic with respect to Haar measure ⟺ T is topologically transitive.

Proof: (⟹) This is obvious as Haar measure is positive on
non-empty open sets.

(⟸) This proof is like the last part of the proof in example (5)
of §4 Chapter 1, which deals with the case when G is connected and
abelian. //

Theorem 5.6:

If T is topologically transitive then T has no nonconstant
continuous invariant functions.

Proof: Suppose f ∈ C(X), fT(x) = f(x). If $\overline{O_T(x_0)}$ = X then f
is constant on $O_T(x_0)$, a dense set, and hence is constant on X. //

Remarks:

(1) If all the T-invariant continuous functions are constant then T
need not be topologically transitive. The following is an example to
illustrate this:

$$X = K^2 \times \{0\} \cup K^2 \times \{1\} \Big/ (e,0) \approx (e,1)$$

i.e., two copies of the two-torus joined at the identity. Let
A: K^2 → K^2 be an ergodic automorphism and define T: X → X by

$$T(x,0) = (Ax,0), \quad T(x,1) = (Ax,1).$$

Then T is not topologically transitive since T preserves $K^2 \times \{0\}$
and $K^2 \times \{1\}$. However, each continuous T-invariant function is constant since it must be constant on both $K^2 \times \{0\}$ and $K^2 \times \{1\}$, because A is ergodic, and these two constants must be the same because
they must agree at the point (e,0) ≡ (e,1). //

(2) T can be topologically transitive and have a dense set of periodic points. To illustrate this we prove that any automorphism A of K^2 has a dense set of periodic points.

Fix n > 0. Consider the finite subgroup of K^2 consisting of points of the form (w_1, w_2) where $w_1^n = w_2^n = 1$. These are all the elements of K^2 of group order n. Since A is an automorphism it preserves this finite subgroup and hence, each member of this subgroup is a periodic point for A. If we now vary n we obtain a dense set. This proof can obviously be extended to an automorphism of K^n, n > 1. //

(3) Topologically transitive homeomorphisms enjoy some of the properties of minimal homeomorphisms and also allow other interesting things to occur; e.g., a dense set of periodic points. (2) and (3) of Theorem 5.4 show that topological transitivity is (in some sense) a topological analogue of ergodicity. Also, topologically transitive homeomorphisms are "indecomposable"; i.e., we cannot write

$$X = \bigcup_{\alpha} E_\alpha, \quad TE_\alpha = E_\alpha \quad \text{and} \quad E_\alpha \quad \text{closed}$$

when T is topologically transitive. So it seems that topologically transitive homeomorphisms are better building blocks than minimal homeomorphisms. If T has a decomposition into minimal pieces then each piece is also topologically transitive. So, the best thing to do is to try to get a decomposition of T into topologically transitive pieces, and then see if these pieces are also minimal.

A distal homeomorphism (i.e., $x \neq y \Rightarrow \exists \; \delta > 0 \; \ni$ $d(T^n(x), T^n(y)) > \delta \; \forall \; n \in Z$) can be decomposed into minimal pieces (Ellis [1]). An Axiom A* homeomorphism can be decomposed into topologically transitive pieces (Smale [1]). But, not all homeomorphisms can be decomposed into topologically transitive pieces; e.g., see the example in remark (1) above.

The following gives a sufficient but not necessary condition for a topologically transitive homeomorphism to be minimal.

Theorem 5.7:

If X is a compact metrizable space, $T: X \to X$ a topologically transitive homeomorphism, and if there exists a metric on X making T an isometry, then T is minimal.

Proof: Suppose such a metric is d, i.e., $d(Tx,Ty) = d(x,y)$. Let $\overline{O_T(x_0)} = X$ and consider $x \in X$. We want to show that $\overline{O_T(x)} = X$. Let $y \in X$ and let $\varepsilon > 0$. There exist $n,m \in Z$ such that

$$d(x,T^m(x_0)) < \varepsilon, \quad d(y,T^n(x_0)) < \varepsilon$$

so,
$$d(y,T^{n-m}(x)) \leq d(y,T^n(x_0)) + d(T^n(x_0),T^{n-m}(x))$$
$$= d(y,T^n(x_0)) + d(T^m(x_0),x)$$
$$< 2\varepsilon.$$

Therefore $\overline{O_T(x)} = X$. //

We now check our examples for topological transitivity.

Examples:

(i) I is topologically transitive iff $X =$ one point.

(ii) $T(x) = ax$ is topologically transitive iff T is minimal iff T is ergodic iff $\{a^n : n \in Z\}$ is dense in X.

Proof: Suppose $\overline{O_T(x_0)} = X$, i.e., the closure of the set $\{a^n x_0 : n \in Z\}$ equals X. There exist $\{n_i\}$ such that

$$a^{n_i} \cdot x_0 \to y \cdot x_0 \quad \text{i.e.,} \quad a^{n_i} \to y.$$

So, $\{a^n : n \in Z\}$ is dense in X. (Another proof would be to apply Theorem 5.7 or Corollary 5.5.) //

(iii) An automorphism A of a compact metric group is topologically transitive iff A is ergodic. (See Corollary 5.5.)

(iv) An affine transformation T of a compact metric group X is topologically transitive iff T is ergodic. (See Corollary 5.5.)

(v) The shift on k symbols is topologically transitive. Consider
$X = \prod_{-\infty}^{\infty} \{0,1,\ldots,k-1\}$, T = shift. We know T is an automorphism of
the compact metric group X. The Haar measure on X is the measure
given by the weights $1/k,\ldots,1/k$. (To check this, fix $x \in X$ and
show, by checking on rectangles and using Theorem 1.1, that this
measure is invariant under translation by x.) T is ergodic with
respect to Haar measure, and therefore T is topologically transitive
by Corollary 5.5.

§3. Topological Conjugacy and Discrete Spectrum

When should we consider two homeomorphisms of compact spaces to
be the "same" from a dynamical point of view? The following seems
the most suitable:

Definition 5.5:

Let T: X → X, S: Y → Y be homeomorphisms of compact spaces.
T is topologically conjugate to S if there exists a homeomorphism
φ: X → Y such that φT = Sφ.

Notes:

(1) This is an equivalence relation.

(2) If T and S are topologically conjugate then T is minimal
iff S is minimal and T is topologically transitive iff S is
topologically transitive.

Definition 5.6:

Let X be a compact metric space, T: X → X a homeomorphism,
f a complex-valued continuous function on X. We say that f is an

<u>eigenfunction</u> for T if ∃ λ ∈ C ∋

$$f(Tx) = \lambda f(x) \quad \forall \ x \in X, \quad \text{and} \quad f \neq 0.$$

We then call λ the corresponding <u>eigenvalue</u> for f.

<u>Remarks</u>:

Suppose T is topologically transitive.

(1) f(Tx) = λf(x), f ∈ C(X) ⟹ |λ| = 1 and |f(x)| = constant.

Proof: f(Tx) = λf(x) ⟹ |f(Tx)| = |λ||f(x)|. Therefore,

$$\sup_{x \in X} |f(Tx)| = |\lambda| \sup_{x \in X} |f(x)|$$

and since T is onto

$$= \sup_{x \in X} |f(x)|.$$

Therefore |λ| = 1. Hence, |f(Tx)| = |f(x)| and by Theorem 5.6 |f(x)| = constant. //

(2) If fT = λf, gT = λg, f,g ∈ C(X) then f = constant·g.

Proof: By (1), |g(x)| ≠ 0 ∀ x ∈ X since g ≠ 0. Therefore (f/g)(Tx) = (f/g)(x) ⟹ f/g = constant by Theorem 5.6. //

(3) Eigenfunctions corresponding to distinct eigenvalues are linearly independent in C(X).

Proof: Let $f_n(Tx) = \lambda_n f_n(x)$ where $\{\lambda_n\}$ are all distinct for n = 1,...,k. Suppose ∀ x ∈ X,

$$a_1 f_1(x) + a_2 f_2(x) + \ldots + a_k f_k(x) = 0$$

where the a_i ∈ C for i = 1,...,k.

By applying the above equation to $T^i x$ instead of x, we get

$$a_1 \lambda_1^i f_1(x) + a_2 \lambda_2^i f_2(x) + \ldots + a_k \lambda_k^i f_k(x) = 0 \quad \forall \ x \in X.$$

Hence

$$
\begin{pmatrix}
1 & 1 & \cdots & 1 \\
\lambda_1 & \lambda_2 & \cdots & \lambda_k \\
\vdots & & & \vdots \\
\lambda_1^{k-1} & \lambda_2^{k-1} & \cdots & \lambda_k^{k-1}
\end{pmatrix}
\begin{pmatrix}
a_1 f_1(x) \\
a_2 f_2(x) \\
\vdots \\
a_k f_k(x)
\end{pmatrix}
=
\begin{pmatrix}
0 \\
0 \\
\vdots \\
0
\end{pmatrix} .
$$

All the λ_i's are distinct so the matrix is nonsingular. Therefore

$$
\begin{pmatrix}
a_1 f_1(x) \\
\vdots \\
a_k f_k(x)
\end{pmatrix}
=
\begin{pmatrix}
0 \\
\vdots \\
0
\end{pmatrix}
\qquad \forall \ x \in X,
$$

i.e., $a_i f_i(x) = 0 \ \forall \ x \in X$, $i = 1,\ldots,k$, i.e., $a_i = 0$, $i = 1,\ldots,k$ since $f_i \neq 0$. Hence, the f_i's are linearly independent in $C(X)$. //

(4) The eigenvalues form a subgroup of the circle group K.

Under our assumptions T has only countably many eigenvalues. To check there are only countably many eigenvalues it suffices to show that if $h: X \to K$ is an eigenfunction corresponding to an eigenvalue $\tau \neq 1$ then $\|h - 1\| > 1/4$. For then two eigenfunctions, with values in K, corresponding to different eigenvalues will be greater than distance $1/4$ apart and, since $C(X)$ has a countable dense set, there can only be countably many eigenvalues. So let $h(Tx) = \tau h(x)$, $\tau \neq 1$. Choose $x_0 \in X$ and p so that $\tau^p h(x_0)$ is in the left-hand half of the unit circle. Then

$$
\|h - 1\| = \sup_{x \in X} \|h(x) - 1\|
$$

$$
\geq \|h(T^p x_0) - 1\| = \|\tau^p h(x_0) - 1\| > 1/4. \qquad //
$$

Definition 5.7:

Let $T: X \to X$ be a homeomorphism of the compact metric space X.

We say that T has <u>topological</u> <u>discrete</u> <u>spectrum</u> if the smallest closed linear subspace of C(X) containing the eigenfunctions of T is C(X), i.e., the eigenfunctions span C(X).

Note:

When T is topologically transitive and has topological discrete spectrum, $\exists$ f_n: X $\rightarrow$ X, n = 1,2,... linearly independent, spanning C(X), such that $f_n T(x) = \lambda_n f_n(x)$. The following is a representation theorem for topologically transitive homeomorphisms with topological discrete spectrum.

Theorem 5.8: (Halmos and von Neumann [1])

The following are equivalent for a homeomorphism T of a compact metric space X:

(1) T is topologically transitive and is an isometry for some metric on X.

(2) T is topologically conjugate to a minimal rotation on a compact abelian metric group.

(3) T is minimal and has topological discrete spectrum.

(4) T is topologically transitive and has topological discrete spectrum.

Proof: (1) $\Rightarrow$ (2). Let d be the isometry metric for T. Suppose $\overline{O_T(x_0)}$ = X. Define a multiplication * in $O_T(x_0)$ by $T^n x_0 * T^m x_0 = T^{n+m} x_0$. We have

$$d(T^n x_0 * T^m x_0, T^p x_0 * T^q x_0) = d(T^{n+m} x_0, T^{p+q} x_0)$$

$$\leq d(T^{n+m} x_0, T^{p+m} x_0) + d(T^{p+m} x_0, T^{p+q} x_0)$$

$$= d(T^n x_0, T^p x_0) + d(T^m x_0, T^q x_0).$$

Hence, the map *: $O_T(x_0) \times O_T(x_0) \rightarrow O_T(x_0)$ is uniformly continuous and therefore can be extended uniquely to a continuous map *: X×X $\rightarrow$ X.

Also, $d(T^{-n}x_0, T^{-m}x_0) = d(T^m x_0, T^n x_0)$ and so,

$$O_T(x_0) \xrightarrow{\text{inverse}} O_T(x_0)$$

is uniformly continuous and can be uniquely extended to a continuous map of X. Thus we get that X is a topological group and is also abelian since it has a dense abelian subgroup $\{T^n x_0 : n \in Z\}$. Since $T(T^n x_0) = T^{n+1} x_0 = T x_0 * T^n x_0$ we have $Tx = Tx_0 * x$ and so T is the rotation by Tx_0.

(2) ⟹ (3). If T is a minimal rotation on a compact abelian group G then each character of G is an eigenfunction. Let A be the collection of all finite linear combinations of characters. Then A is a subalgebra of C(X), contains the constants, is closed under complex conjugation, and separates points. Applying the Stone-Weierstrass Theorem we see that $\overline{A} = C(X)$.

(3) ⟹ (4) is trivial.

(4) ⟹ (1). We can choose eigenfunctions $f_n : X \to K$, $n \geq 1$, with $f_n(T) = \lambda_n f_n$ and where the f_n are linearly independent and span C(X). Define a metric on X by:

$$d(x,y) = \sum_{n=1}^{\infty} \frac{|f_n(x) - f_n(y)|}{2^n} .$$

Then $\qquad d(Tx, Ty) = \sum_{n=1}^{\infty} \frac{|\lambda_n f_n(x) - \lambda_n f_n(y)|}{2^n} = d(x,y).$

It remains to check that d gives the topology on X. If $d(x_n, x) \to 0$ then for all $n \geq 1$, as $m \to \infty$,

$$\frac{1}{2^n} |f_n(x_m) - f_n(x)| \leq d(x_m, x) \to 0.$$

Thus, $\forall n \geq 1$, $f_n(x_m) \to f_n(x)$ as $m \to \infty$ and since $\{f_n\}$ separates points, $x_m \to x$ as $m \to \infty$. Conversely, suppose $x_m \to x$. Let $\varepsilon > 0$, and choose N such that

$$\sum_{n=N+1}^{\infty} \frac{2}{2^n} < \frac{\varepsilon}{2} .$$

By the continuity of the functions $f_1, \ldots, f_N$ $\exists$ M $\ni$ $m > M$ $\Rightarrow$ $|f_i(x_m) - f_i(x)| < \varepsilon/2$ $i = 1, \ldots, N$. If $m > M$ then

$$d(x_m, x) = \sum_{i=1}^{\infty} \frac{1}{2^i} |f_i(x_m) - f_i(x)|$$

$$\leq \sum_{i=1}^{N} \frac{1}{2^i} \cdot \frac{\varepsilon}{2} + \frac{\varepsilon}{2} \leq \varepsilon$$

i.e., $d(x_m, x) \to 0$. //

Remark:

If $Tx = ax$ is a minimal rotation of a compact metric abelian group G it is straightforward to show that the set of eigenvalues of T is $\{\gamma(a): \gamma \in \hat{G}\}$ and every eigenfunction is a constant multiple of a character. In fact, this follows from Theorem 3.3 since each continuous eigenfunction is an L^2-eigenfunction.

We have the following isomorphism theorem.

Theorem 5.9: (Topological Discrete Spectrum Theorem)

Two minimal homeomorphisms of compact metric spaces both having topological discrete spectrum are topologically conjugate iff they have the same eigenvalues.

Proof: (1) The proof is along the lines of the proof of Theorem 3.1, but instead of using Theorem 2.1 we use the Banach-Stone Theorem. This says that if X, Y are compact spaces, $\Phi: C(Y) \to C(X)$ is a bijective linear isometry, and $\Phi(f \cdot g) = \Phi(f)\Phi(g)$, then there exists a homeomorphism $\phi: X \to Y$ such that $\Phi(f)(x) = f(\phi(x))$.

(2) This theorem can also be proved using Theorem 5.8 and character theory. By Theorem 5.8 we can suppose T is a minimal rotation of a compact abelian group G, Tx = ax, and S is a minimal rotation of a compact abelian group H, Sy = by. We are assuming $\{\gamma(a): \gamma \in \hat{G}\} = \{\delta(b): \delta \in \hat{H}\}$. Define a map $\theta: \hat{H} \to \hat{G}$ by $\theta(\delta)(a) = \delta(b)$. This is well-defined and a bijection. Moreover, θ is easily checked to be a group automorphism and hence induces an automorphism C: G → H. It is easy to show that CT = SC. //

Remark:

Thus the theory of topologically transitive homeomorphisms with topological discrete spectrum is entirely analogous to that of ergodic measure-preserving transformations with pure point spectrum.

§4. Invariant Measures for Homeomorphisms

In this section we consider some connections between the topological and measure theoretic systems. We first prove some results about Borel measures including the fact that a Borel measure on a metric space is determined by how it integrates continuous functions. By a Borel measure on X is meant a measure defined on the Borel subsets of X, (i.e., the smallest σ-algebra containing the closed sets).

Theorem 5.10:

A Borel probability measure m on a metric space X is regular (i.e., if $\mathcal{B}$ denotes the Borel sets then $\forall$ B $\in \mathcal{B}$ and $\forall$ $\varepsilon > 0$ $\exists$ an open set U_ε and a closed set C_ε with $C_\varepsilon \subseteq B \subseteq U_\varepsilon$ and $m(U_\varepsilon \setminus C_\varepsilon) < \varepsilon$).

Proof: (The proof does not require X to be metric but that each closed set be a G_δ.) Let R be the collection of all sets such that the regularity condition holds, i.e., $R = \{A \in \mathcal{B}: \forall \varepsilon > 0 \exists$ open U_ε, closed C_ε with $C_\varepsilon \subseteq A \subseteq U_\varepsilon$ and $m(U_\varepsilon \setminus C_\varepsilon) < \varepsilon\}$. We show that R is a σ-algebra. Let $A \in R$; we show that $X \backslash A \in R$. Let $\varepsilon > 0$. $\exists$ open U_ε, closed C_ε with $C_\varepsilon \subseteq A \subseteq U_\varepsilon$ $\ni$ $m(U_\varepsilon \backslash C_\varepsilon) < \varepsilon$. Thus, $X \backslash U_\varepsilon \subseteq X \backslash A \subseteq X \backslash C_\varepsilon$ and $(X \backslash C_\varepsilon) \backslash (X \backslash U_\varepsilon) = U_\varepsilon \backslash C_\varepsilon$, so

$$m((X \backslash C_\varepsilon) \backslash (X \backslash U_\varepsilon)) = m(U_\varepsilon \backslash C_\varepsilon) < \varepsilon.$$

Therefore $X \backslash A \in R$.

We now show R is closed under countable unions. Let $A_1, A_2, \ldots \in R$ and let $A = \bigcup_{i=1}^{\infty} A_i$. Let $\varepsilon > 0$ be given. There exist open $U_{\varepsilon,n}$, closed $C_{\varepsilon,n}$ such that $C_{\varepsilon,n} \subseteq A_n \subseteq U_{\varepsilon,n}$ and $m(U_{\varepsilon,n} \backslash C_{\varepsilon,n}) < \varepsilon/3^n$. Let $U_\varepsilon = \bigcup_{n=1}^{\infty} U_{\varepsilon,n}$ (which is open), $\tilde{C}_\varepsilon = \bigcup_{n=1}^{\infty} C_{\varepsilon,n}$, and choose k such that $m(\tilde{C}_\varepsilon \backslash \bigcup_{n=1}^{k} C_{\varepsilon,n}) < \varepsilon/2$. Let $C_\varepsilon = \bigcup_{n=1}^{k} C_{\varepsilon,n}$ (which is closed). We have $C_\varepsilon \subseteq A \subseteq U_\varepsilon$. Also,

$$m(U_\varepsilon \backslash C_\varepsilon) \leq m(U_\varepsilon \backslash \tilde{C}_\varepsilon) + m(\tilde{C}_\varepsilon \backslash C_\varepsilon)$$

$$\leq \sum_{n=1}^{\infty} m(U_{\varepsilon,n} \backslash C_{\varepsilon,n}) + m(\tilde{C}_\varepsilon \backslash C_\varepsilon)$$

$$\leq \sum_{n=1}^{\infty} \frac{\varepsilon}{3^n} + \frac{\varepsilon}{2} = \varepsilon.$$

Therefore R is a σ-algebra.

To complete the proof we show that R contains all the closed subsets of X. Let C be a closed set and $\varepsilon > 0$. Define $U_n = \{x \in X: d(C,x) < 1/n\}$. This is an open set, $U_1 \supseteq U_2 \supseteq \ldots \supseteq U_n \supseteq \ldots$ and $\bigcap_{i=1}^{\infty} U_i = C$. Choose k such that $m(U_k \backslash C) < \varepsilon$ and let $C_\varepsilon = C$ and $U_\varepsilon = U_k$. This shows $C \in R$. //

Corollary 5.10:

For a Borel probability measure m on a metric space X we have
that for a Berel set B

$$m(B) = \sup_{\substack{C \text{ closed} \\ C \subseteq B}} m(C) \quad \text{and} \quad m(B) = \inf_{\substack{U \text{ open} \\ U \supseteq B}} m(U).$$

Theorem 5.11:

Let m,μ be two Borel probability measures on the metric
space X. If $\int_X f\ dm = \int_X f\ d\mu$ ∀ f ∈ C(X) then m = μ.

Proof: By the above corollary it suffices to show that m(C) =
μ(C) for all closed sets C ⊆ X. Suppose C is closed and let
ε > 0. By the regularity of m there exists an open set U, U ⊇ C
such that m(U\C) < ε.

Define f: X → R by

$$f(x) = \begin{cases} 0 & \text{if } x \notin U \\ \dfrac{d(x,X\backslash U)}{d(x,X\backslash U) + d(x,C)} & \text{if } x \in U. \end{cases}$$

f is well-defined since the denominator is not zero. Also f is
continuous, f = 0 on X\U, f = 1 on C, and 0 ≤ f(x) ≤ 1 ∀
x ∈ X. Then,

$$\mu(C) \le \int_X f\ d\mu = \int_X f\ dm \le m(U) < m(C) + \varepsilon.$$

Therefore μ(C) < m(C) + ε ∀ ε > 0, so μ(C) ≤ m(C). By symmetry
we get that m(C) ≤ μ(C). //

Theorem 5.12: (Riesz Representation Theorem)

Let X be a compact metric space and J: C(X) → C a continuous
linear map such that J is a positive operator (i.e., if f ≥ 0 then
J(f) ≥ 0) and J(1) = 1. Then there exists a Borel probability

measure μ on X such that

$$J(f) = \int_X f \, d\mu$$

for all f in C(X).

Proof: See Halmos [1], p. 247. //

The next theorem expresses the fact that the unit ball in the dual space of C(X) is weakly compact.

Theorem 5.13:

If $\{\mu_n\}$ is a sequence of Borel probability measures on a compact metric space X, then there is a subsequence $\{\mu_n\}$ which converges in the weak topology, i.e., $\exists$ a Borel probability measure μ on X such that

$$\int_X f \, d\mu_{n_i} \to \int_X f \, d\mu$$

for all f in C(X).

Proof: We write $\mu(f) = \int_X f \, d\mu$ when $f \in C(X)$ and μ is a Borel measure. Choose $f_1, f_2, \ldots$ dense in C(X). Consider the sequence of complex numbers $\{\mu_n(f_1)\}$. This is bounded by $\|f_1\|$, and so has a convergent subsequence, say $\{\mu_n^{(1)}(f_1)\}$. Consider $\{\mu_n^{(1)}(f_2)\}$; this is bounded and so has a convergent subsequence $\{\mu_n^{(2)}(f_2)\}$. Notice that $\{\mu_n^{(2)}(f_1)\}$ also converges. We proceed in this manner, and for each $i \geq 1$, construct a subsequence $\{\mu_n^{(i)}\}$ of $\{\mu_n\}$ such that $\{\mu_n^{(i)}\} \subseteq \{\mu_n^{(i-1)}\} \subseteq \ldots \subseteq \{\mu_n^{(1)}\} \subseteq \{\mu_n\}$, and so that $\{\mu_n^{(i)}(f)\}$ converges for $f = f_1, f_2, \ldots, f_i$. Consider the diagonal $\{\mu_n^{(n)}\}$. The sequence $\{\mu_n^{(n)}(f_i)\}$ converges for all i; thus $\{\mu_n^{(n)}(f)\}$ converges for all $f \in C(X)$ (by an easy approximation argument). Let $J(f) = \lim\limits_{n\to\infty} \mu_n^{(n)}(f)$. Clearly $J: C(X) \to C$ is linear and bounded, as $|J(f)| \leq \|f\|$. Also $J(1) = 1$, and if $f \geq 0$ then $J(f) \geq 0$. By the Riesz Theorem, there exists a Borel probability measure μ on X such that $J(f) = \int_X f\, d\mu$ for all $f \in C(X)$, i.e.,

$$\int_X f\, d\mu_n^{(n)} \to \int_X f\, d\mu. \quad //$$

Corollary 5.13:

The space of Borel probability measures on a compact metric space X is itself a compact metric space under the weak topology.

Proof: Let $f_1, f_2, \ldots$ be dense in $C(X)$. Define

$$D(m,\mu) = \sum_{i=1}^{\infty} \frac{\left| \int_X f_i\, dm - \int_X f_i\, d\mu \right|}{2^i \|f_i\|}.$$

D is a metric on the space of Borel probability measures which gives rise to the weak topology. The compactness follows from the previous theorem. //

Theorem 5.14: (Krylov and Bogolioubov [1])

If T is a homeomorphism of a compact metric space X then there exists a Borel probability measure on X which is preserved by T.

Proof: Fix $x \in X$. For $f \in C(X)$ and $n \geq 0$, define

$$J_n(f) = \frac{1}{n} \sum_{i=0}^{n-1} f(T^i(x)).$$

$J_n: C(X) \to C$ satisfies the conditions of the Riesz Representation Theorem (note that $|J_n(f)| \leq \|f\|$), so there exists a Borel probability measure μ_n on X such that

$$J_n(f) = \int_X f \, d\mu_n \quad \text{for all} \ f \in C(X).$$

By Theorem 5.13 there exists a subsequence $\{\mu_{n_j}\}$ and a Borel probability measure μ on X such that

$$J_{n_j}(f) = \int_X f \, d\mu_{n_j} \ \to \ \int_X f \, d\mu \quad \text{for all} \ f \in C(X).$$

Since

$$|J_{n_j}(f \circ T) - J_{n_j}(f)| = \frac{1}{n_j} |fT^{n_j}(x) - f(x)|$$

$$\leq \frac{1}{n_j} \cdot 2\|f\| \ \to \ 0 \quad \text{as} \ j \to \infty$$

we have

$$\int_X f \circ T \, d\mu = \int_X f \, d\mu,$$

i.e.,

$$\int_X f \, d\mu T^{-1} = \int_X f \, d\mu \quad \forall \ f \in C(X).$$

So by the uniqueness Theorem 5.11 for Borel measures we have that $\mu(T^{-1}B) = \mu(B)$ for all Borel sets B. //

Theorem 5.15:

Let T be a homeomorphism of a compact metric space X, and let M_T denote the collection of all T-invariant Borel probability measures on X (by Theorem 5.14, $M_T \neq \phi$). Then

(1) M_T is closed in the weak topology,

(2) M_T is a convex set, and

(3) if $m \in M_T$ then m is an extreme point of M_T iff m is ergodic with respect to T.

Proof: (1) Suppose $\{\mu_n\} \subset M_T$ converges to μ in the weak topology. Then

$$\int fT \, d\mu_n \;\rightarrow\; \int fT \, d\mu$$

$$\|$$

$$\int f \, d\mu_n \;\rightarrow\; \int f \, d\mu$$

so that μ is T-invariant.

(2) is obvious.

(3) Suppose $m \in M_T$, m not ergodic. There exists a Borel set E such that $T^{-1}E = E$ a.e. and $0 < m(E) < 1$. Define measures m_1 and m_2 by

$$m_1(B) = \frac{m(B \cap E)}{m(E)} \quad \text{and} \quad m_2(B) = \frac{m(B \cap (X\backslash E))}{m(X\backslash E)}$$

for B a Borel set. Note that m_1 and m_2 are in M_T, $m_1 \neq m_2$, and

$$m(B) = m(E)m_1(B) + (1 - m(E))m_2(B),$$

so that m is not an extreme point of M_T.

Conversely, suppose $m \in M_T$ is ergodic, and

$$m = \alpha m_1 + (1-\alpha)m_2$$

where $m_1, m_2 \in M_T$, $0 \le \alpha \le 1$. We must show $m_1 = m_2$. $m_1 \ll m$ (m_1 is absolutely continuous with respect to m) so that the Radon-Nikodym derivative dm_1/dm exists, (i.e., $m_1(E) = \int_E \dfrac{dm_1(x)}{dm} dm(x)$).

So,
$$m_1(E) = m_1(T^{-1}E) = \int_{T^{-1}E} \frac{dm_1}{dm} \, dm(x)$$

$$= \int_E \frac{dm_1}{dm}(T^{-1}y) \, dm(T^{-1}y)$$

$$= \int_E \frac{dm_1}{dm}(T^{-1}y) \, dm(y).$$

Therefore $\frac{dm_1}{dm}(T^{-1}y) = \frac{dm_1}{dm}(y)$ a.e.(m) (by uniqueness of the Radon-Nikodym derivative). But, m is ergodic, so that $dm_1/dm = $ constant = k a.e.(m). Therefore

$$1 = m_1(X) = \int_X k \, dm = k \cdot m(X) = k.$$

Since k = 1, $m_1 = m$ and therefore $m_2 = m = m_1$. //

Definition 5.8:

T is <u>uniquely ergodic</u> if there is only one T-invariant Borel probability measure on X, i.e., M_T = one point.

Remark:

T is uniquely ergodic with respect to m implies that T is ergodic with respect to m.

Unique ergodicity is connected to minimality by:

Theorem 5.16:

Suppose T is uniquely ergodic and m is its unique invariant measure. T is minimal iff m(U) > 0 for all nonempty open sets U.

Proof: Suppose T is minimal. If U is open, U ≠ φ then $X = \bigcup_{n=-\infty}^{\infty} T^n(U)$; so if m(U) = 0 then m(X) = 0, a contradiction.

Conversely, suppose m(U) > 0 for all open nonempty U. Suppose also that T is not minimal, i.e., there exists a closed set K such that TK = K, K ≠ X. $T|_K$ has an invariant Borel probability measure μ_K on K by Theorem 5.14. Define μ on X by μ(B) = $\mu_K(K \cap B)$ for all Borel sets B. Then $\mu \in M_T$ and μ ≠ m because m(X\K) > 0 as X\K is nonempty and open while μ(X\K) = 0. This contradicts the unique ergodicity of T. //

The following results formulate unique ergodicity in terms of ergodic averages.

Theorem 5.17:

The following are equivalent:

(1) ∀ f ∈ C(X), $\frac{1}{n}\sum\limits_{i=0}^{n-1} f(T^i x)$ converges uniformly to a constant.

(2) ∀ f ∈ C(X), $\frac{1}{n}\sum\limits_{i=0}^{n-1} f(T^i x)$ converges pointwise on X to a

constant.

(3) ∃ m ∈ M_T ∋ ∀ f ∈ C(X) and ∀ x ∈ X,

$$\frac{1}{n}\sum_{i=0}^{n-1} f(T^i x) \;\rightarrow\; \int f\ dm.$$

(4) T is uniquely ergodic.

Proof: (1) ⇒ (2) holds trivially.

(2) ⇒ (3). Define k: C(X) → C by

$$k(f) = \lim_{n\to\infty} \frac{1}{n}\sum_{i=0}^{n-1} fT^i(x).$$

Observe that k is a linear operator and is continuous since

$$\Big|\ \frac{1}{n}\sum_{i=0}^{n-1} fT^i(x)\ \Big| \;\le\; |f|.$$

Also k(1) = 1 and f ≥ 0 ⇒ k(f) ≥ 0. Thus by the Riesz Representation Theorem there exists a Borel probability measure m such that
k(f) = $\int$ f dm. But k(fT) = k(f) and so, $\int$ fT dm = $\int$ f dm, i.e.,
$\int$ f dmT^{-1} = $\int$ f dm which implies that mT^{-1} = m by 5.11, so that
m ∈ M_T.

(3) ⇒ (4). Suppose that ν ∈ M_T. We have

$$\frac{1}{n}\sum_{i=0}^{n-1} fT^i(x) \;\rightarrow\; f^* \qquad ∀\ x,$$

where $f^* = \int f\ dm$. Integrating with respect to ν, and using the bounded convergence theorem we get that

$$\int f \, d\nu = \int f^* \, d\nu = f^* = \int f \, dm \quad \forall \quad f \in C(X).$$

Hence $\nu = m$ by 5.11. Therefore T is uniquely ergodic.

(4) $\Rightarrow$ (1). If $\frac{1}{n} \sum_{i=0}^{n-1} fT^i(x)$ converges uniformly to a constant

then this constant must be $\int f \, dm$, where m is the unique T-invariant

measure. Suppose (1) is false. Then $\exists \, g \in C(X)$, $\exists \, \varepsilon > 0 \ni \forall \, N$

$\exists \, n > N$ and $\exists \, x_n \in X \ni$

$$| \frac{1}{n} \sum_{i=0}^{n-1} gT^i(x_n) - \int g \, dm \, | \geq \varepsilon \, .$$

Define $J_n \colon C(X) \to C$ by $J_n(f) = \frac{1}{n} \sum_{i=0}^{n-1} fT^i(x_n)$. J_n satisfies the

conditions of the Riesz Representation Theorem. Hence, $J_n(f) = \int f \, d\mu_n$

for some Borel probability measure μ_n. Moreover, there exists a sub-

sequence $\{\mu_{n_i}\}$ such that

$$J_{n_i}(f) = \int f \, d\mu_{n_i} \quad \to \quad \int f \, d\mu$$

for all $f \in C(X)$ and for some Borel probability measure μ (by

Theorem 5.13). Then

$$|J_{n_i}(fT) - J_{n_i}(f)| = \frac{1}{n_i} |fT^{n_i}(x_{n_i}) - f(x_{n_i})|$$

$$\leq \frac{2\|f\|}{n_i} \to 0,$$

so that $$\int fT \, d\mu = \int f \, d\mu.$$

Hence, $\mu \in M_T$. But, $|\int g \, d\mu - \int g \, dm| \geq \varepsilon$ so that $\mu \neq m$ contra-

dicting the uniqueness of m. //

We now see which of our examples are uniquely ergodic.

Examples:

(i)   I is uniquely ergodic iff X = one point, since M_T = all
Borel probability measures.

(ii)   T(x) = ax on a compact group is uniquely ergodic iff T is
minimal.

     Proof: (⇒) follows from Theorem 5.16 and the fact that Haar
measure is positive on open sets.

     (⇐) T is minimal ⟷ $\{a^n\}$ is dense in G. Therefore G is
abelian. If $1 \neq \gamma \in \hat{G}$ then

$$\frac{1}{n} \sum_{i=0}^{n-1} \gamma(T^i x) = \frac{1}{n} \sum_{i=0}^{n-1} \gamma(a^i)\gamma(x) = \frac{\gamma(x)}{n} \frac{(\gamma(a)^n - 1)}{\gamma(a) - 1} \rightarrow 0 \quad \text{as} \quad n \rightarrow \infty$$

(note that $\gamma(a) \neq 1$). So (2) of Theorem 5.17 holds when f is a
character and the condition (2) will hold for each $f \in C(X)$ by
approximation, since finite linear combinations of characters are
dense in C(X). //

(iii) An automorphism of a compact group G is uniquely ergodic iff
G = {e}, since Haar measure is preserved and so is the point measure
concentrated at e.

(iv)   An affine transformation of a compact connected abelian metric
group is uniquely ergodic iff it is minimal.

     Proof: (⇒) follows by Theorem 5.16.

     (⇐) follows, as in example (ii), by checking that (2) of Theo-
rem 5.17 holds. This was done by Hahn and Parry [1]. //

(v)   The Bernoulli shift on k symbols is uniquely ergodic iff
k = 1. This is by example (iii).

    An excellent survey of unique ergodicity and related topics can
be found in J. C. Oxtoby [1].

Recent results of Jewett [1] and Krieger [1] imply that any ergodic invertible measure-preserving transformation of a Lebesgue space is isomorphic in the sense of Chapter 2 to a uniquely ergodic system. This indicates a certain lack of measure-theoretic import for the concept of unique ergodicity. Hahn and Katznelson [1] have found uniquely ergodic transformations in shift spaces with arbitrarily large measure-theoretic entropies.

Chapter 6: Topological Entropy

Adler, Konheim, and McAndrew [1] introduced topological entropy as an invariant of topological conjugacy and also as an analogue of measure theoretic entropy.

§1. Definition by Open Covers

All logarithms are to base 2. Let X be a compact topological space. We shall be interested in open covers of X which we denote by $\alpha, \beta, \ldots$.

Definition 6.1:

If α, β are open covers of X their join $\alpha \vee \beta$ is given by: $\alpha \vee \beta = \{A \cap B : A \in \alpha, B \in \beta\}$.

Definition 6.2:

An open cover β is a refinement of an open cover α, written $\alpha < \beta$, if every member of β is a subset of a member of α. In particular, $\alpha < \alpha \vee \beta$, $\beta < \alpha \vee \beta$.

Definition 6.3:

If α is an open cover of X and $T: X \to X$ is continuous then $T^{-1}\alpha = \{T^{-1}(A) : A \in \alpha\}$ is an open cover of X.

Note:

$$T^{-1}(\alpha \vee \beta) = T^{-1}(\alpha) \vee T^{-1}(\beta) \quad \text{and} \quad \alpha < \beta \implies T^{-1}\alpha < T^{-1}\beta.$$

Definition 6.4:

If α is an open cover of X let $N(\alpha)$ = the number of sets in a finite subcover of α with smallest cardinality. We define the entropy of α by: $H(\alpha) = \log N(\alpha)$.

Remarks:

(1) $H(\alpha) \geq 0$.

(2) $H(\alpha) = 0$ iff $N(\alpha) = 1$ iff $X \in \alpha$. $H(\alpha)$ is small means that there are a few sets in α which cover X. $H(\alpha)$ is large means that some part of X is covered by a large number of sets in α and not by a small number.

(3) $\alpha < \beta \Rightarrow H(\alpha) \leq H(\beta)$.

  Proof: Let $\{B_1, \ldots, B_{N(\beta)}\}$ be a subcover of β with minimal cardinality. $\forall$ i $\exists$ $A_i \in \alpha$ $\ni$ $A_i \supseteq B_i$. So, $\{A_1, \ldots, A_{N(\beta)}\}$ covers X and is a subcover of α. Thus $N(\alpha) \leq N(\beta)$. //

(4) $H(\alpha \vee \beta) \leq H(\alpha) + H(\beta)$.

  Proof: Let $\{A_1, \ldots, A_{N(\alpha)}\}$ be a subcover of α of minimal cardinality, and $\{B_1, \ldots, B_{N(\beta)}\}$ be a subcover of β of minimal cardinality. Then

$$\{A_i \cap B_j : \ 1 \leq i \leq N(\alpha), \ 1 \leq j \leq N(\beta)\}$$

is a subcover of $\alpha \vee \beta$ so, $N(\alpha \vee \beta) \leq N(\alpha)N(\beta)$. //

(5) If $T: X \to X$ is a continuous map then $H(T^{-1}\alpha) \leq H(\alpha)$. If T is also surjective then $H(T^{-1}\alpha) = H(\alpha)$.

  Proof: If $\{A_1, \ldots, A_{N(\alpha)}\}$ is a subcover of α of minimal cardinality then $\{T^{-1}A_1, \ldots, T^{-1}A_{N(\alpha)}\}$ is a subcover of $T^{-1}\alpha$, so $N(T^{-1}\alpha) \leq N(\alpha)$. If T is onto, and $\{T^{-1}A_1, \ldots, T^{-1}A_{N(T^{-1}\alpha)}\}$ is a subcover of $T^{-1}\alpha$ of minimal cardinality then $\{A_1, \ldots, A_{N(T^{-1}\alpha)}\}$ also covers X, so $N(\alpha) \leq N(T^{-1}\alpha)$. //

Theorem 6.1:

  If α is an open cover of X and $T: X \to X$ is continuous, then

$$\lim_{n \to \infty} \frac{1}{n} H(\alpha \vee T^{-1}\alpha \vee \ldots \vee T^{-(n-1)}\alpha) \quad \text{exists.}$$

Proof: Recall that if we set

$$a_n = H(\alpha \vee T^{-1}\alpha \vee \ldots \vee T^{-(n-1)}\alpha)$$

then by Theorem 4.4 it suffices to show that:

$$a_n \geq 0, \quad \text{and} \quad a_{n+m} \leq a_n + a_m \quad \forall \ m,n.$$

By (1), $a_n \geq 0$, and

$$a_{n+m} = H(\alpha \vee T^{-1}\alpha \vee \ldots \vee T^{-(n+m-1)}\alpha)$$

$$\leq H(\alpha \vee T^{-1}\alpha \vee \ldots \vee T^{-(n-1)}\alpha)$$

$$+ H(T^{-n}\alpha \vee \ldots \vee T^{-(n+m-1)}\alpha) \quad \text{by (4)}$$

$$= a_n + H(T^{-n}\alpha \vee \ldots \vee T^{-(n+m-1)}\alpha)$$

$$= a_n + H(T^{-n}(\alpha \vee \ldots \vee T^{-(m-1)}\alpha))$$

$$\leq a_n + H(\alpha \vee \ldots \vee T^{-(m-1)}\alpha) \quad \text{by (5)}$$

$$= a_n + a_m. \quad //$$

Definition 6.5:

If α is an open cover of X and $T: X \to X$ is a continuous map then the __entropy__ of T __relative to__ α is given by:

$$h(T,\alpha) = \lim_{n\to\infty} \frac{1}{n} H(\alpha \vee T^{-1}\alpha \vee \ldots \vee T^{-(n-1)}\alpha).$$

Remarks:

(6) $h(T,\alpha) \geq 0$ by (1).

(7) $\alpha < \beta \Rightarrow h(T,\alpha) \leq h(T,\beta)$.

Proof: $\alpha < \beta \Rightarrow \bigvee_{i=0}^{n-1} T^{-i}\alpha < \bigvee_{i=0}^{n-1} T^{-i}\beta$, so by (3) we have that $H(\bigvee_{i=0}^{n-1} T^{-i}\alpha) \leq H(\bigvee_{i=0}^{n-1} T^{-i}\beta)$. Hence $h(T,\alpha) \leq h(T,\beta)$. //

Note that if β is a finite subcover of α then $\alpha < \beta$ so then $h(T,\alpha) \leq h(T,\beta)$.

(8) $h(T,\alpha) \leq H(\alpha)$.

Proof: By (4) we have

$$H(\alpha \vee T^{-1}\alpha \vee \ldots \vee T^{-(n-1)}\alpha)$$

$$\leq \sum_{i=0}^{n-1} H(T^{-i}\alpha)$$

$$\leq n \cdot H(\alpha) \quad \text{by (5)}. \quad //$$

Definition 6.6:

If $T: X \to X$ is continuous, the topological entropy of T is given by:

$$h(T) = \sup_\alpha h(T,\alpha)$$

where α ranges over all open covers of X.

Remarks:

(9) $h(T) \geq 0$.

(10) In the definition of $h(T)$ one can take the supremum over finite covers of X. This follows from (7).

(11) $h(I) = 0$ where I is the identity map of X.

The next result shows that topological entropy is an invariant of topological conjugacy.

Theorem 6.2:

If X_1, X_2 are compact spaces and $T_i: X_i \to X_i$ are continuous for $i = 1,2$, and are topologically conjugate, then they have the same entropy.

Proof: Suppose $\phi: X_1 \to X_2$ is a homeomorphism such that $\phi T_1 = T_2 \phi$. Let α be an open cover of X_2. Then,

$$h(T_2, \alpha) = \lim_n \frac{1}{n} H(\alpha \vee T_2^{-1}\alpha \vee \ldots \vee T_2^{-(n-1)}\alpha)$$

$$= \lim_n \frac{1}{n} H(\phi^{-1}(\alpha \vee T_2^{-1}\alpha \vee \ldots \vee T_2^{-(n-1)}\alpha)) \quad \text{by (5)}$$

$$= \lim_n \frac{1}{n} H(\phi^{-1}\alpha \vee T_1^{-1}\phi^{-1}\alpha \vee \ldots \vee T_1^{-(n-1)}\phi^{-1}\alpha)$$

$$= h(T_1, \phi^{-1}\alpha).$$

By taking suprema the result follows. //

Adler, Konheim, and McAndrew proved several results about $h(T)$. In the next section we give a definition of topological entropy for any uniformly continuous map of a metric space (not necessarily compact). This definition can also be given for uniform spaces. The definition will reduce to the previous definition in the compact case. We shall prove the properties of $h(T)$ with this new definition. However, one result we would like to note is the following:

Theorem 6.3:

If $T: X \rightarrow X$ is a homeomorphism of a compact space X, then $h(T) = h(T^{-1})$.

Proof:

$$h(T, \alpha) = \lim_n \frac{1}{n} H(\alpha \vee T^{-1}\alpha \vee \ldots \vee T^{-(n-1)}\alpha)$$

$$= \lim_n \frac{1}{n} H(T^{n-1}(\alpha \vee T^{-1}\alpha \vee \ldots \vee T^{-(n-1)}\alpha))$$

$$= \lim_n \frac{1}{n} H(\alpha \vee T\alpha \vee \ldots \vee T^{n-1}\alpha)$$

$$= h(T^{-1}, \alpha). \quad //$$

Adler, Konheim, and McAndrew could not answer the following questions which they stated as conjectures in their paper.

Notation:

From now on the measure theoretic entropy of a transformation T preserving a measure m will be written $h_m(T)$.

Conjecture 1:

Let X be compact and m a regular Borel measure on X. If T: X → X is a homeomorphism preserving m then

$$h_m(T) \leq h(T).$$

[This has been proved by Goodwyn [1]. T need only be continuous. We shall give a proof when X is a finite-dimensional torus. (See Theorem 6.9.]

Conjecture 2:

Let X be a compact metric space and T: X → X a homeomorphism. (By Theorems 5.10 and 5.14 we know that M_T, the set of Borel measures on X invariant under T , is nonempty.) Then $h(T) = \sup_{m \in M_T} h_m(T)$.

[Partial contributions were made by Goodwyn [1] and Dinaburg [1], but Goodman [1] finally proved it. We can drop the condition of X being metric and allow T to be only continuous, provided we define M_T to be all T-invariant regular Borel probability measures on X.]

Conjecture 3:

If X is a compact metric group and T is an automorphism of X then $h(T) = h_m(T)$ where m is Haar measure.

[This was shown by Berg [1] and generalized by Bowen [4]: T can be an affine transformation and need not necessarily be invertible. We shall prove this. (See Theorem 6.10.)]

Conjecture 4:

Let $\{T_t\}$ be a one-parameter group of homeomorphisms of a compact space X. Then $h(T_t) = |t| h(T_1)$.

[This was proved by Bowen when X is metric.]

Conjecture 5:

Let X,Y be compact spaces. Let $\{T_x: x \in X\}$ be a family of homeomorphisms of Y so that

$$T(x,y) = (x,T_x(y))$$

is a continuous map of X×Y. Then

$$h(T) = \sup_{x \in X} h(T_x).$$

[This was proved in the case where X and Y are metric by Bowen.]

§2. Bowen's Definition

If (X,d) is a metric space and $x \in X$ then $B_\varepsilon(x)$ will denote the open ball centered at x and of radius ε. UC(X,d) will denote the collection of all uniformly continuous maps $\phi: X \to X$.

Let $T \in UC(X,d)$; $n \in Z$, n > 0; and $\varepsilon > 0$. If $K \subseteq X$, a set $F \subseteq X$ is said to (n,ε)-span K with respect to T if $\forall$ $x \in K$ $\exists$ $y \in F$ such that

$$\max_{0 \le i \le n-1} d(T^i x, T^i y) \le \varepsilon.$$

For K compact, let $r_n(\varepsilon,K)$ be the smallest cardinality of any (n,ε)-spanning set for K with respect to T. We show later (Theorem 6.4) that $r_n(\varepsilon,K) < \infty$.

Set $$\bar{r}_T(\varepsilon,K) = \limsup_n \frac{1}{n} \log r_n(\varepsilon,K).$$

A set $E \subset X$ is (n,ε)-separated with respect to T if, whenever x,y $\in$ E, $x \ne y$ then

$$\max_{0 \le i \le n-1} d(T^i x, T^i y) > \varepsilon.$$

For K compact, let $s_n(\varepsilon,K)$ denote the largest cardinality of any (n,ε)-separated subset of K with respect to T. We later show (Theorem 6.4) that $s_n(\varepsilon,K)$ is finite.

Set $\qquad\qquad \bar{s}_T(\varepsilon,K) = \lim \sup \frac{1}{n} \log s_n(\varepsilon,K).$

We define $h(T,K) = \lim_{\varepsilon\to0} \bar{r}_T(\varepsilon,K) = \lim_{\varepsilon\to0} \bar{s}_T(\varepsilon,K).$ These limits exist and are equal by Theorem 6.4. We then define

$$h_d(T) = \sup_{K \text{ compact}} h(T,K).$$

Remarks:

(1) This definition can also be given in the context of uniform spaces.

(2) $h_d(T)$ measures the amount of expansion in T (for the metric d). For $r_n(\varepsilon,K)$ and $s_n(\varepsilon,K)$ to increase as n increases we need some expansion for T.

(3) The ideas for this definition come from the work of Kolmogorov on the size of a metric space. If (X,ρ) is a metric space then a subset F is said to ε-span X if $\forall$ x $\in$ X $\exists$ y $\in$ F with $\rho(x,y) \le \varepsilon$, and a subset E is said to be ε-separated if whenever y,z $\in$ E, y $\ne$ z, then $\rho(y,z) > \varepsilon$. The ε-entropy of (X,ρ) is then the logarithm of the minimum number of elements of an ε-spanning set and the ε-capacity is the logarithm of the maximum number of elements in an ε-separated set. So in the above definitions we are considering the metric spaces (K,d_n) where d_n is the restriction to the compact set K of the metric

$$\rho_n(x,y) = \max_{0\le i\le n-1} d(T^i x, T^i y).$$

Then $h(T,K) = \lim_{\varepsilon\to0} \lim_{n\to\infty} \sup \frac{1}{n} [\varepsilon\text{-entropy of } (K,\rho_n)].$ (It follows from the proof of the next theorem that to define h(T,K) it suffices to consider spanning sets for K which are subsets of K.) $\bar{s}_T(\varepsilon,K)$ is the average ε-capacity of the spaces (K,d_n) and h(T,K) is the limit of $\bar{s}_T(\varepsilon,K)$ as $\varepsilon \to 0$.

<u>Theorem</u> 6.4:

   Suppose K is compact. Then

(i) $r_n(\varepsilon,K) \leq s_n(\varepsilon,K) \leq r_n(\varepsilon/2,K) < \infty$ and

(ii) if $\varepsilon_1 < \varepsilon_2$ then $\bar{r}_T(\varepsilon_1,K) \geq \bar{r}_T(\varepsilon_2,K)$ and

$$\bar{s}_T(\varepsilon_1,K) \geq \bar{s}_T(\varepsilon_2,K).$$

   <u>Proof</u>: (i). We first show $r_n(\varepsilon,K) < \infty$. There exists a $\delta > 0$ such that $d(x,y) < \delta$ implies

$$\max_{0 \leq i \leq n-1} d(T^i x, T^i y) < \varepsilon.$$

Then $r_n(\varepsilon,K)$ is less than the number of δ-balls needed to cover K and hence is finite.

   We shall now prove $s_n(\varepsilon,K) \leq r_n(\varepsilon/2,K)$. Suppose $E \subseteq K$ is an (n,ε)-separated set and that F $(n,\varepsilon/2)$-spans K. Define $\phi : E \to F$ by choosing for each $x \in E$ some point $\phi(x) \in F$ with $\max_{0 \leq i \leq n-1} d(T^i \phi(x), T^i(x)) \leq \varepsilon/2$. If $\phi(x) = \phi(y)$ then

$$\max_{0 \leq i \leq n-1} d(T^i x, T^i y) \leq \varepsilon/2 + \varepsilon/2 = \varepsilon$$

so that x = y. Hence ϕ is one-to-one and the cardinality of E is less than or equal to the cardinality of F. Therefore $s_n(\varepsilon,K) \leq r_n(\varepsilon/2,K)$.

   Finally we show $r_n(\varepsilon,K) \leq s_n(\varepsilon,K)$. Let E be an (n,ε)-separated subset of K of maximum cardinality. We claim that E (n,ε)-spans K, since if not $\exists$ $x \in K$ $\ni$

$$\max_{0 \leq i \leq n-1} d(T^i x, T^i y) > \varepsilon \quad \forall \ y \in E.$$

Then $E \cup \{x\}$ is an (n,ε)-separated subset of K, contradicting the choice of E.

(ii) is obvious. //

Hence the definition of $h(T,K)$ makes sense.

Remarks:

(1) $h_d(T)$ depends on d.

(2) If $K \subseteq K_1 \cup \ldots \cup K_m$ are all compact then

$$h(T,K) \leq \max_{1 \leq i \leq m} h(T,K_i).$$

Proof: Certainly, $s_n(\varepsilon,K) \leq s_n(\varepsilon,K_1) + \ldots + s_n(\varepsilon,K_m)$. Fix $\varepsilon > 0$. $\forall$ n choose $K_{i_n}(\varepsilon)$ $\ni$

$$s_n(\varepsilon,K_{i_n}(\varepsilon)) = \max_j s_n(\varepsilon,K_j).$$

Then $s_n(\varepsilon,K) \leq m \cdot s_n(\varepsilon,K_{i_n}(\varepsilon))$ and so,

$$\log s_n(\varepsilon,K) \leq \log m + \log s_n(\varepsilon,K_{i_n}(\varepsilon)).$$

Choose $n_j \to \infty$ such that

$$\frac{1}{n_j} \log s_{n_j}(\varepsilon,K) \to \limsup \frac{1}{n} \log s_n(\varepsilon,K)$$

and so that $K_{i_{n_j}}(\varepsilon)$ does not depend on j (i.e., $K_{i_{n_j}}(\varepsilon) = K(\varepsilon)$ $\forall$ j). Thus, $\bar{s}_T(\varepsilon,K) \leq \bar{s}_T(\varepsilon,K(\varepsilon))$.

Choose $\varepsilon_\alpha \to 0$ so that $K(\varepsilon_\alpha)$ is constant ($= K_{i_0}$, say). Thus,

$h(T,K) \leq h(T,K_{i_0}) \leq \max_j h(T,K_j).$ //

(3) $\forall$ $\delta > 0$, in order to compute $h_d(T)$ it suffices to take the supremum of $h(T,K)$ over compact sets of diameter less than δ. This is true by (2).

(4) If X is compact, $h_d(T) = h(T,X)$.

Proof: By (2), if $K \subset X$, K compact, then

$$h(T,K) \leq h(T,X). \quad //$$

Definition 6.7:

Two metrics d and d' on X are uniformly equivalent if

$$\text{id.: } (X,d) \rightarrow (X,d') \quad \text{and}$$
$$\text{id.: } (X,d') \rightarrow (X,d)$$

are both uniformly continuous.

In this case, $T \in UC(X,d)$ iff $T \in UC(X,d')$.

Theorem 6.5:

If d and d' are uniformly equivalent and $T \in UC(X,d)$ then $h_d(T) = h_{d'}(T)$.

Proof: Let $\varepsilon_1 > 0$. Choose $\varepsilon_2 > 0$ $\ni$

$$d'(x,y) < \varepsilon_2 \quad \Rightarrow \quad d(x,y) < \varepsilon_1$$

and choose $\varepsilon_3 > 0$ $\ni$

$$d(x,y) < \varepsilon_3 \quad \Rightarrow \quad d'(x,y) < \varepsilon_2.$$

Let K be compact. Then

$$r_n(\varepsilon_1,K,d) \leq r_n(\varepsilon_2,K,d') \quad \text{and}$$
$$r_n(\varepsilon_2,K,d') \leq r_n(\varepsilon_3,K,d).$$

Hence, $\qquad \bar{r}_T(\varepsilon_1,K,d) \leq \bar{r}_T(\varepsilon_2,K,d') \leq \bar{r}_T(\varepsilon_3,K,d).$

If $\varepsilon_1 \to 0$, then $\varepsilon_2 \to 0$, and $\varepsilon_3 \to 0$ so we have

$$h_d(T,K) = h_{d'}(T,K). \quad //$$

Remark:

If X is compact and if d and d' are equivalent metrics then they are uniformly equivalent. Also, each continuous map $T: X \to X$ is uniformly continuous.

Theorem 6.6: (Lebesgue Covering Lemma)

If (X,d) is a compact metric space and α is a finite open cover of X then there exists a $\delta > 0$ such that each subset of X of diameter $\leq \delta$ lies in some member of α.

Proof: Let $\alpha = \{A_1,\ldots,A_p\}$. Assume the theorem is false. Then for all n there exists $B_n \subseteq X$ such that $\mathrm{diam}(B_n) \leq 1/n$ and B_n is not contained in any A_i. Choose $x_n \in B_n$ and select a subsequence $\{x_{n_i}\}$ which converges, say $x_{n_i} \to x$. Suppose $x \in A_j \in \alpha$. Let $a = d(x,X\backslash A_j) > 0$. Choose n_i such that $n_i > 2/a$ and $d(x_{n_i},x) < a/2$. Then if $y \in B_{n_i}$

$$d(y,x) \leq d(y,x_{n_i}) + d(x_{n_i},x) \leq \frac{1}{n_i} + \frac{a}{2} < a.$$

So $y \in A_j$. Hence $B_{n_i} \subseteq A_j$, a contradiction. $//$

Theorem 6.7:

When X is compact, Bowen's definition of entropy coincides with the open cover definition.

Proof: For the duration of this proof let $h^*(T,\alpha)$ and $h^*(T)$ denote the numbers that occur in the open cover definition.

Let $\alpha = \{A_1,\ldots,A_p\}$ be an open cover of X. We shall show that $h^*(T,\alpha) \leq h(T)$. Let δ be a Lebesgue number for α. Let F be a

$(n,\delta/2)$-spanning set for X of minimum cardinality. For $z \in F$ choose $A_{i_0}(z), \ldots, A_{i_{n-1}}(z)$ in α so that $\overline{B_{\delta/2}(T^k z)} \subseteq A_{i_k}(z)$. Let

$$C(z) \equiv A_{i_0}(z) \cap T^{-1} A_{i_1}(z) \cap \ldots \cap T^{-(n-1)} A_{i_{n-1}}(z),$$

which is a member of $\alpha \vee T^{-1}\alpha \vee \ldots \vee T^{-(n-1)}\alpha$.

We have $X = \bigcup_{z \in F} C(z)$ since if $x \in X \; \exists \; z \in F \; \ni$

$$\max_{0 \le i \le n-1} d(T^i x, T^i z) \le \delta/2$$

and hence $x \in T^{-k}(B_{\delta/2}(T^k z)) \subseteq T^{-k} A_{i_k}(z)$, $0 \le k \le n-1$; so $x \in C(z)$.

Hence $N(\alpha \vee T^{-1}\alpha \vee \ldots \vee T^{-(n-1)}\alpha) \le |F| = r_n(\delta/2, X)$, and

$$h^*(T,\alpha) \le \bar{r}_T(\delta/2, X) \le h(T, X) = h(T)$$

since X is compact. Therefore $h^*(T) \le h(T)$.

To prove the converse let $\delta > 0$ be given. Choose an open cover $\alpha = \{A_1, \ldots, A_p\}$ of X such that $\text{diam}(A_i) < \delta$ for all i. Let E be an (n,δ)-separated subset of X with maximal cardinality. Two members of E cannot belong to the same element of $\alpha \vee T^{-1}\alpha \vee \ldots \vee T^{-(n-1)}\alpha$ since if

$$x,y \in \bigcap_{j=0}^{n-1} T^{-j} A_{i_j} \qquad x,y \in E$$

then $\max_{0 \le j \le n-1} d(T^j x, T^j y) < \delta$ and so $x = y$.

So, $\quad N(\alpha \vee T^{-1}\alpha \vee \ldots \vee T^{-(n-1)}\alpha) \ge |E| = s_n(\delta, X)$.

Therefore $\quad h^*(T) \ge h^*(T,\alpha) \ge \bar{s}_T(\delta, X)$.

Letting $\delta \to 0$ we have $h^*(T) \ge h(T, X) = h(T)$. //

Notes:

(1) If we had set up the definitions using uniformities we would get the above for compact Hausdorff spaces.

(2) Since a maximal separated set is spanning we get by the first part of the proof of Theorem 6.7 that $s_n(\delta/2,X) \geq N(\alpha \vee T^{-1}\alpha \vee \ldots \vee T^{-(n-1)}\alpha)$ where δ is a Lebesgue number for α.

Theorem 6.8:

(1) If $T \in UC(X,d)$ and $m > 0$ then $h(T^m) = m \cdot h(T)$.

(2) Let $T_i \in UC(X_i,d_i)$ $i = 1,2$. Define a metric on $X_1 \times X_2$ by $d((x_1,x_2),(y_1,y_2)) = \max \{d_1(x_1,y_1),d_2(x_2,y_2)\}$. Then

$$h_d(T_1 \times T_2) \leq h_{d_1}(T_1) + h_{d_2}(T_2).$$

If X_1 and X_2 are compact then equality holds.

   Proof: (1). Since $r_n(\varepsilon,K,T^m) \leq r_{mn}(\varepsilon,K,T)$ we have

$$\frac{1}{n} \log r_n(\varepsilon,K,T^m) \leq \frac{m}{mn} \log r_{mn}(\varepsilon,K,T)$$

and therefore $h_d(T^m) \leq m \cdot h_d(T)$.

   Since T is uniformly continuous, $\forall\ \varepsilon > 0\ \exists\ \delta > 0\ \ni$

$$d(x,y) < \delta \quad \Rightarrow \quad \max_{0 \leq j \leq m-1} d(T^j x, T^j y) < \varepsilon.$$

So an (n,δ)-spanning set for K with respect to T^m is also an (nm,ε)-spanning set for K with respect to T. Hence, $r_n(\delta,K,T^m) \geq r_{mn}(\varepsilon,K,T)$ so, $m \cdot \bar{r}_T(\varepsilon,K) \leq \bar{r}_{T^m}(\delta,K)$. Thus,

$$m \cdot h_d(T,K) \leq h_d(T^m,K).$$

   (2). Let $K_i \subseteq X_i$ be compact, $i = 1,2$. If F_i is an (n,ε)-spanning set for K_i with respect to T_i then $F_1 \times F_2$ is an

(n,ε)-spanning set for $K_1 \times K_2$ with respect to $T_1 \times T_2$. Hence,

$$r_n(\varepsilon, K_1 \times K_2, T_1 \times T_2) \leq r_n(\varepsilon, K_1, T_1) \cdot r_n(\varepsilon, K_2, T_2)$$

which implies

$$\bar{r}_{T_1 \times T_2}(\varepsilon, K_1 \times K_2) \leq \bar{r}_{T_1}(\varepsilon, K_1) + \bar{r}_{T_2}(\varepsilon, K_2).$$

Therefore

$$h_d(T_1 \times T_2, K_1 \times K_2) \leq h_{d_1}(T_1, K_1) + h_{d_2}(T_2, K_2).$$

Let $\pi_i : X_1 \times X_2 \to X_i$ $i = 1, 2$ be the projection map. If $K \subseteq X_1 \times X_2$ is compact then $K_1 = \pi_1(K)$ and $K_2 = \pi_2(K)$ are compact and $K \subseteq K_1 \times K_2$. Hence

$$h_d(T_1 \times T_2, K) \leq h_d(T_1 \times T_2, K_1 \times K_2).$$

Therefore

$$h_d(T_1 \times T_2) = \sup_{\substack{K \subseteq X_1 \times X_2 \\ \text{compact}}} h_d(T_1 \times T_2, K)$$

$$= \sup_{\substack{K_1 \subseteq X_1 \\ K_2 \subseteq X_2 \\ \text{cpt.}}} h_d(T_1 \times T_2, K_1 \times K_2)$$

$$\leq \sup_{\substack{K_1 \subseteq X_1 \\ \text{cpt.}}} h_{d_1}(T_1, K_1) + \sup_{\substack{K_2 \subseteq X_2 \\ \text{cpt.}}} h_{d_2}(T_2, K_2)$$

$$= h_{d_1}(T_1) + h_{d_2}(T_2).$$

Now suppose X_1 and X_2 are compact. Let α_i be an open cover of X_i and have Lebesgue number δ_i $(i = 1, 2)$. If S_i is a maximal $(n, \delta_i/2)$-separated set for X_i with respect to T_i then $S_1 \times S_2$ is an (n, δ)-separated set for $X_1 \times X_2$ with respect to $T_1 \times T_2$ where

$\delta = \min(\delta_1/2, \delta_2/2)$. Therefore

$$s_n(\delta, X_1 \times X_2) \geq s_n(\delta_1/2, X_1) \cdot s_n(\delta_2/2, X_2)$$

$$\geq N(\alpha_1 \vee T_1^{-1}\alpha_1 \vee \ldots \vee T_1^{-(n-1)}\alpha_1) \cdot N(\alpha_2 \vee T_2^{-1}\alpha_2 \vee \ldots \vee T_2^{-(n-1)}\alpha_2)$$

by note 2 above. Hence

$$h(T_1 \times T_2) \geq \limsup_{n \to \infty} \frac{1}{n} \log s_n(\delta, X_1 \times X_2)$$

$$\geq \lim_{n \to \infty} \frac{1}{n} \log N(\alpha_1 \vee T_1^{-1}\alpha_1 \vee \ldots \vee T_1^{-(n-1)}\alpha_1) \; +$$

$$\lim_{n \to \infty} \frac{1}{n} \log N(\alpha_2 \vee T_2^{-1}\alpha_2 \vee \ldots \vee T_2^{-(n-1)}\alpha_2)$$

$$= h(T_1, \alpha_1) + h(T_2, \alpha_2).$$

Since α_1, α_2 were arbitrary we get $h(T_1 \times T_2) \geq h(T_1) + h(T_2)$. //

Remarks:

(1) If T is a homeomorphism and $T \in UC(X,d)$, $T^{-1} \in UC(X,d)$ then $h_d(T) \neq h_d(T^{-1})$ in general. We shall show later that if $T: R \to R$ is defined by $T(x) = 2x$ then $h(T) = \log 2$ while $h(T^{-1}) = 0$ using the usual metric on R. (Note that T has expansion but T^{-1} does not.) However, on compact spaces $h(T) = h(T^{-1})$ (Theorem 6.3). This is because on a compact space T^{-1} has "as much expansion" as does T.

(2) Equality probably holds in (2) for non-compact X_1 and X_2 but I do not know a proof.

§3. Connections with Measure Theoretic Entropy

In this section we shall prove conjecture 3 (assuming conjecture 1 is true) and we shall prove conjecture 1 when X is a finite-dimensional torus.

Theorem 6.9: (Goodwyn)

Let X be a compact space and $T: X \to X$ continuous. If m is a T-invariant regular Borel probability measure on X, then $h_m(T) \leq h(T)$.

We shall prove this theorem when X is a finite-dimensional torus since the proof is easier in this case.

Proof: Let $X = K^k$, $T: K^k \to K^k$ be any continuous map, and let m be any T-invariant Borel probability measure on K^k. We wish to show that $h_m(T) \leq h(T)$.

Consider K^k as R^k/Z^k with metric

$$d(x+Z^k, y+Z^k) = \inf_{v \in Z^k} \| x - y + v \| \qquad x,y \in R^k$$

where $\| \cdot \|$ denotes the usual Euclidean norm.

Fix an integer $q \geq 0$. Consider a decomposition of the unit k-cube in R^k into all sets of the form

$$\left\{ (x_1, \ldots, x_k): \frac{p_1}{2^q} \leq x_1 < \frac{p_1+1}{2^q}, \quad \frac{p_2}{2^q} \leq x_2 < \frac{p_2+1}{2^q}, \right.$$

$$\ldots, \quad \frac{p_k}{2^q} \leq x_k < \frac{p_k+1}{2^q} \quad \text{where}$$

$$\left. 0 \leq p_i < 2^q \text{ for } i = 1, \ldots, k \right\}.$$

This induces a partition of the torus K^k which we denote by $\xi_q = \{A_1, \ldots, A_{2^{kq}}\}$. Any ball in K^k of radius $\frac{1}{2^{q+2}}$ intersects at most 2^k members of ξ_q. Let $\gamma = \{C_1, \ldots, C_s\}$ be a cover of K^k by open balls of radius $\frac{1}{2^{q+2}}$. Let 2δ be a Lebesgue number for γ. For all $x \in K^k$, fix some $C(x) \in \gamma$ with $B_\delta(x) \subseteq C(x)$. Let F be

an (n,δ)-spanning set for K^k with respect to T of minimal cardi-
nality. Let $x \in \bigcap\limits_{j=0}^{n-1} T^{-j}A_{i_j}$, $A_{i_j} \in \xi_q$. Choose $y \in F$ with

$$\max_{0 \le j \le n-1} d(T^j x, T^j y) \le \delta,$$

and hence, $T^j x \in C(T^j y)$. Thus $T^j x \in A_{i_j} \cap C(T^j y)$. Hence if

$$U_n = \{(i_0, \ldots, i_{n-1}): \bigcap\limits_{j=0}^{n-1} T^{-j}A_{i_j} \ne \phi\}$$

then $|U_n| \le 2^{kn}|F| = 2^{kn} r_n(\delta, K^k)$. So, using Corollary 4.2,

$$H_m(\xi_q \vee T^{-1}\xi_q \vee \ldots \vee T^{-(n-1)}\xi_q) \le \log |U_n| \le n \cdot \log 2^k + \log r_n(\delta, K^k).$$

Thus,
$$h_m(T, \xi_q) \le \log 2^k + \overline{\lim_n} \frac{1}{n} \log r_n(\delta, K^k)$$

$$= \log 2^k + \bar{r}_T(\delta, K^k)$$

$$\le \log 2^k + h(T) \quad = \quad k + h(T).$$

(Note that we are taking logarithms to base 2.)
But, $A(\xi_q) \nearrow B$ as $q \to \infty$. So, by Theorem 4.14

$$h_m(T) = \lim_{q \to \infty} h_m(T, \xi_q) \le k + h(T).$$

But this holds for any continuous T, so, in particular, for T^n,
$n > 0$. If $n > 0$

$$h_m(T) = \frac{1}{n} h_m(T^n) \le \frac{1}{n} [k + h(T^n)] = \frac{k}{n} + h(T)$$

so, by letting $n \to \infty$ we get the desired result. //

Theorem 6.10: (Bowen)

Let X be a compact metric group and $T: X \to X$, $T = a \cdot A$ an affine transformation. If m denotes Haar measure on X then $h_m(T) = h(T) = h_m(A) = h(A)$.

Proof: By the previous theorem $h_m(T) \le h(T)$, and so it remains to prove that $h(T) \le h_m(T)$. Suppose d is a left invariant metric on X. Let $B_\varepsilon(y) = \{x: d(x,y) < \varepsilon\}$ and

$$D_n(x,\varepsilon,T) = \bigcap_{k=0}^{n-1} T^{-k}B_\varepsilon(T^kx).$$

By induction we shall show that

$$T^{-k}B_\varepsilon(T^kx) = x \cdot (A^{-k}B_\varepsilon(e)).$$

It is true for $k = 0$ by the invariance of the metric d. Assuming it holds for k we prove it for $k+1$.

$$T^{-(k+1)}B_\varepsilon(T^{k+1}x) = T^{-1}(T^{-k}B_\varepsilon(T^k(Tx)))$$

$$= T^{-1}(Tx \cdot A^{-k}B_\varepsilon(e))$$

$$= x \cdot (A^{-(k+1)}B_\varepsilon(e)).$$

Hence,
$$D_n(x,\varepsilon,T) = x \cdot \bigcap_{k=0}^{n-1} A^{-k}B_\varepsilon(e) = x \cdot D_n(e,\varepsilon,A).$$

Also,
$$m(D_n(x,\varepsilon,T)) = m(D_n(e,\varepsilon,A)).$$

Let $\varepsilon > 0$. Let $\zeta = \{A_1,\dots,A_n\}$ be a partition of X into Borel sets of diameter $< \varepsilon$. If $x \in \bigcap_{k=0}^{n-1} T^{-k}A_{i_k}$ then $\bigcap_{k=0}^{n-1} T^{-k}A_{i_k} \subseteq$

$x \cdot D_n(e,\varepsilon,A)$, since if $y \in \bigcap_{k=0}^{n-1} T^{-k}A_{i_k}$ then $T^k(x), T^k(y) \in A_{i_k}$ $\forall$ k,

and hence $y \in T^{-k}B_\varepsilon(T^kx)$ $\forall$ k, i.e., $y \in D_n(x,\varepsilon,T) = x \cdot D_n(e,\varepsilon,A)$.

Thus, $m(\bigcap_{k=0}^{n-1} T^{-k}A_{i_k}) \leq m(D_n(e,\varepsilon,A))$ and taking logs we see that

$$\sum_{i_0,\ldots,i_{n-1}=1}^{n} m(\bigcap T^{-k}A_{i_k}) \log m(\bigcap T^{-k}A_{i_k})$$

$$\leq \sum_{i_0,\ldots,i_{n-1}=1}^{n} m(\bigcap T^{-k}A_{i_k}) \log m(D_n(e,\varepsilon,A))$$

$$= \log m(D_n(e,\varepsilon,A)).$$

Thus, $h_m(T) \geq h_m(T,\zeta) = \lim_n \frac{1}{n} H(\zeta \vee \ldots \vee T^{-(n-1)}\zeta)$

$$\geq \limsup_n [-\frac{1}{n} \log m(D_n(e,\varepsilon,A))].$$

Hence, since ε was arbitrary we obtain that

$$h_m(T) \geq \lim_{\varepsilon \to 0} \limsup_n [-\frac{1}{n} \log m(D_n(e,\varepsilon,A))].$$

(The limit clearly exists.) Consider now an (n,ε)-separated set with respect to T, $E \subseteq X$, having maximal cardinality. Then

$$\bigcup_{x \in E} D_n(x,\varepsilon/2,T) = \bigcup_{x \in E} x \cdot D_n(e,\varepsilon/2,A)$$

is a disjoint union because of the choice of E. Therefore

$$s_n(\varepsilon,X) \cdot m(D_n(e,\varepsilon/2,A)) \leq 1$$

and so $$s_n(\varepsilon,X) \leq \frac{1}{m(D_n(e,\varepsilon/2,A))}.$$

Therefore $$\bar{s}_T(\varepsilon,X) \leq \limsup_n [-\frac{1}{n} \log m(D_n(e,\varepsilon/2,A))]$$

and letting $\varepsilon \to 0$ we see that

$$h(T) = h_d(T,X) \leq \lim_{\varepsilon \to 0} \limsup_{n} \left[- \frac{1}{n} \log m(D_n(e, \varepsilon/2, A))\right]$$

$$\leq h_m(T).$$

Thus, $$h_m(T) = h(T) = \lim_{\varepsilon \to 0} \limsup_{n} \left[- \frac{1}{n} \log m(D_n(e, \varepsilon, A))\right].$$

We can replace T by A here since the right hand side is independent of a. //

Note:

   The formula

$$h(T) = \lim_{\varepsilon \to 0} \limsup_{n} \left[- \frac{1}{n} \log m(D_n(e, \varepsilon, A))\right]$$

illustrates that T measures "the amount of expansion" in T.

§4. <u>Topological</u> <u>Entropy</u> <u>of</u> <u>Linear</u> <u>Maps</u> <u>and</u> <u>Toral</u> <u>Affines</u>

   Our aim in this section is to compute the topological entropy (and hence by Theorem 6.10 the measure theoretic entropy) of affine transformations of finite-dimensional tori. Recall (see §6 of Chapter 0) that we can view the n-torus K^n either multiplicatively as K×K×...×K (n factors) or additively as R^n/Z^n. Each endomorphism A of K^n onto K^n is given, in the additive notation, by

$$A(x + Z^n) = [A] \cdot x + Z^n \qquad x \in R^n,$$

where [A] is an n×n nonsingular matrix with integer entries. [A] determines a linear transformation $\tilde{A}$ of R^n and $\pi\tilde{A} = A\pi$ where $\pi: R^n \to K^n$ is the natural projection given by $\pi(x) = x + Z^n$.

   Let $\|\cdot\|$ denote the usual Euclidean norm on R^n. We define a metric d on R^n/Z^n by

$$d(x+Z^n, y+Z^n) = \inf_{v \in Z^n} \|x - y + v\| \qquad x, y \in R^n.$$

d is left and right invariant and, for every $x \in R^n$ π maps the ball of radius $1/4$ about x in R^n isometrically onto the ball of radius $1/4$ about $\pi(x)$ in R^n/Z^n.

The next theorem deals with such a situation and asserts that $h_d(A) = h_{\tilde{d}}(\tilde{A})$ in this case, where $\tilde{d}$ denotes the metric on R^n induced by the Euclidean norm $\|\cdot\|$. (Since $\|\tilde{A}x - \tilde{A}y\| \le \|\tilde{A}\| \cdot \|x - y\|$ we know $\tilde{A} \in UC(R^n, \tilde{d})$.) This will reduce the problem of calculating the entropy of A to that of calculating the entropy of $\tilde{A}$.

Theorem 6.11:

Let $(X, d), (\tilde{X}, \tilde{d})$ be metric spaces and $\pi : \tilde{X} \to X$ a continuous surjection such that there exists $\delta > 0$ with

$$\pi|_{B_\delta(\tilde{x})} : B_\delta(\tilde{x}) \to B_\delta(\pi(\tilde{x}))$$

an isometric surjection for all $\tilde{x} \in \tilde{X}$. If $T \in UC(X, d)$ and $\tilde{T} \in UC(\tilde{X}, \tilde{d})$ satisfy $\pi\tilde{T} = T\pi$ then

$$h_d(T) = h_{\tilde{d}}(\tilde{T}).$$

Proof: If $\tilde{K}$ is compact in $\tilde{X}$ of diameter $< \delta$ then $\pi(\tilde{K})$ is compact in X of diameter $< \delta$. Every compact subset of X of diameter $< \delta$ is of this form. Let $\varepsilon > 0$ be such that $\varepsilon < \delta$ and if $\tilde{d}(\tilde{x}, \tilde{y}) < \varepsilon$ then $\tilde{d}(\tilde{T}\tilde{x}, \tilde{T}\tilde{y}) < \delta$.

Suppose $\tilde{E} \subseteq \tilde{K}$ is an (n, ε)-separating set with respect to $\tilde{T}$. We first prove that $\pi(\tilde{E})$ is an (n, ε)-separating subset of $\pi(\tilde{K})$ with respect to T. To prove this, let $\tilde{x} \ne \tilde{y}$ belong to $\tilde{E}$. Then $\pi(\tilde{x}) \ne \pi(\tilde{y})$. Let i_0 be chosen so that $\tilde{d}(\tilde{T}^i\tilde{x}, \tilde{T}^i\tilde{y}) \le \varepsilon$ if $i \le i_0$ and $\tilde{d}(\tilde{T}^{i_0+1}\tilde{x}, \tilde{T}^{i_0+1}\tilde{y}) > \varepsilon$. (This can be done since $\tilde{E}$ is an (n, ε)-separating set with respect to $\tilde{T}$.) By our choice of ε,

$\tilde{d}(\tilde{T}^{i_0+1}\tilde{x}, \tilde{T}^{i_0+1}\tilde{y}) < \delta$ and so $\tilde{T}^{i_0+1}\tilde{y} \in B_\delta(\tilde{T}^{i_0+1}\tilde{x})$ which is mapped isometrically onto $B_\delta(T^{i_0+1}\pi(\tilde{x}))$. So,

$$d(T^{i_0+1}\pi(\tilde{x}), T^{i_0+1}\pi(\tilde{y})) = \tilde{d}(\tilde{T}^{i_0+1}\tilde{x}, \tilde{T}^{i_0+1}\tilde{y}) > \varepsilon.$$

Thus $\pi(E)$ is (n,ε)-separated with respect to T. Therefore,

$$s_n(\varepsilon, \tilde{K}, \tilde{T}) \le s_n(\varepsilon, \pi(\tilde{K}), T).$$

To prove the converse inequality, suppose E is an (n,ε)-separated subset of $\pi(\tilde{K}) \subseteq X$ with respect to T, where $\tilde{K}$ is compact and of diameter $< \delta$. Let $\tilde{E} = \pi^{-1}(E) \cap \tilde{K}$. Then $\tilde{E}$ is an (n,ε)-separated set with respect to $\tilde{T}$ since if $\tilde{d}(\tilde{T}^i\tilde{x}, \tilde{T}^i\tilde{y}) \le \varepsilon$ where $\tilde{x}, \tilde{y} \in \tilde{E}$ then $d(T^i\pi(\tilde{x}), T^i\pi(\tilde{y})) \le \varepsilon$. Hence,

$$s_n(\varepsilon, \pi(\tilde{K}), T) \le s_n(\varepsilon, \tilde{K}, \tilde{T}).$$

Therefore $\qquad\qquad s_n(\varepsilon, \tilde{K}, \tilde{T}) = s_n(\varepsilon, \pi(\tilde{K}), T)$

and hence $\qquad\qquad h_{\tilde{d}}(\tilde{T}, \tilde{K}) = h_d(T, \pi(\tilde{K}))$.

By remark (3) of §2

$$h_{\tilde{d}}(\tilde{T}) = h_d(T). \quad //$$

Corollary 6.11:

If $A: K^n \to K^n$ is an endomorphism then $h_d(A) = h_{\tilde{d}}(\tilde{A})$ where $\tilde{A}$ is the linear map of R^n covering A, $\tilde{d}$ is the metric on R^n determined from the Euclidean norm and d is any metric on K^n.

We shall now proceed towards calculating the entropy of a linear map of R^n.

Theorem 6.12:

Suppose A: $R^p \rightarrow R^p$ is a linear map, and ρ a metric determined by a norm on R^p. Then:

(i) $h_\rho(A) \geq \log |\det A|$ if $\det A \neq 0$,   and

(ii) if all the eigenvalues of A have the same absolute value τ then

$$h_\rho(A) = \max \{0, p \cdot \log \tau\}.$$

Proof: All norms on R^p are uniformly equivalent, so, by Theorem 6.5 we can assume that ρ is the metric given by the Euclidean norm. Obviously, $A \in UC(R^p, \rho)$ as

$$\rho(Ax, Ay) = \|Ax - Ay\| \leq \|A\| \|x - y\| = \|A\| \rho(x, y).$$

(i). Let m denote Lebesgue measure on R^p. Then

$$m(A(E)) = |\det A| \cdot m(E)$$

for all Borel sets $E \subseteq R^p$. Let $K \subseteq R^p$ be compact and $m(K) > 0$. If F (n, ε)-spans K then $K \subseteq \bigcup_{x \in F} \overline{D_n(x, \varepsilon, A)} = \bigcup_{x \in F} [x + \overline{D_n(0, \varepsilon, A)}]$

where $D_n(x, \varepsilon, A) = \bigcap_{i=0}^{n-1} A^{-i} B_\varepsilon(A^i x)$ (as in the proof of Theorem 6.10).

Thus, $m(K) \leq m(\overline{D_n(0, \varepsilon, A)}) \cdot r_n(\varepsilon, K) = m(D_n(0, \varepsilon, A)) \cdot r_n(\varepsilon, K)$

i.e., $r_n(\varepsilon, K) \geq \dfrac{m(K)}{m(D_n(0, \varepsilon, A))}$.

Therefore $\bar{r}_A(\varepsilon, K) \geq \overline{\lim_n} \dfrac{1}{n} [\log m(K) - \log m(D_n(0, \varepsilon, A))]$

$$= \overline{\lim_n} [-\frac{1}{n} \log m(D_n(0, \varepsilon, A))].$$

But, $m(D_n(0, \varepsilon, A)) \leq m(A^{-(n-1)} B_\varepsilon(0)) = \dfrac{m(B_\varepsilon(0))}{|\det A|^{n-1}}$

so that $\bar{r}_A(\varepsilon,K) \geq \overline{\lim_{n}} \frac{1}{n} [\log |\det A|^{n-1} - \log m(B_\varepsilon(0))]$

$$= \log |\det A|.$$

Therefore $h_\rho(A) \geq h(A,K) \geq \log |\det A|.$

(ii). In this case,

$$|\det A| = |\text{product of eigenvalues}| = \tau^p.$$

So, by (i) $h_\rho(A) \geq p \log \tau$ and then, clearly,

$$h_\rho(A) \geq \max \{0, p \log \tau\}.$$

We now have to show the opposite inequality. Assume first that
$\|A\| > 1$. Let K be a compact subset of R^p of diameter $< 1/2$.
Choose $b \in R^p$ such that $0 \in b + K \equiv K_b$. The diameter of $K_b < 1/2$
so that $K_b \subset I_1^p$ where

$$I_1^p \equiv \{(x_1,\ldots,x_p) \in R^p : |x_i| \leq 1 \quad \forall \quad i\}.$$

For δ such that $0 < \delta < 1$ let

$$F(\delta) \equiv \{(n_1\delta,\ldots,n_p\delta): n_i \in Z, \quad |n_i\delta| < 2\}.$$

Observe that $|F(\delta)| \leq (5/\delta)^p$, and $\exists$ $c > 0$ $\ni$ $\forall$ $y \in I_2^p$, $\exists$
$x \in F(\delta)$ $\ni$ $\rho(x,y) < c\delta$. $F(\delta)$ is an $(n,\|A\|^n c\delta)$-spanning set for
K_b with respect to A, since if $y \in K_b$ $\exists$ $x \in F(\delta)$ $\ni$

$$\rho(A^i x, A^i y) \leq \|A^i\|\rho(x,y) \leq \|A\|^i c\delta \leq \|A\|^n c\delta,$$
$$\text{for} \quad 0 \leq i \leq n-1.$$

But then, $F(\delta) - b$ is an $(n,\|A\|^n c\delta)$-spanning set with respect to A
for K. Let $\varepsilon > 0$, and set $\delta = \dfrac{\varepsilon}{\|A\|^n \cdot c} < 1$, for n sufficiently
large. Thus

$$r_n(\varepsilon,K) \le \left| F\left(\frac{\varepsilon}{\|A\|^n c}\right) \right| \le \left(\frac{5\|A\|^n c}{\varepsilon}\right)^p$$

for sufficiently large n. Also,

$$\bar{r}_A(\varepsilon,K) = \overline{\lim_n} \frac{1}{n} \log r_n(\varepsilon,K)$$

$$\le \overline{\lim_n} \frac{p}{n} [\log 5 + n \log \|A\| + \log c - \log \varepsilon]$$

$$= p \log \|A\|.$$

So,
$$h_\rho(A) \le p \log \|A\|$$

and hence
$$h_\rho(A) \le \max \{0, p \log \|A\|\} \quad \text{if} \quad \|A\| > 1.$$

If $\|A\| \le 1$ then $h_\rho(A) = 0$ since a $(1,\varepsilon)$-spanning set is an (n,ε)-spanning set. Thus, in all cases

$$h_\rho(A) \le \max \{0, p \log \|A\|\}.$$

However,
$$h_\rho(A) = \frac{1}{n} h_\rho(A^n) \qquad \text{for} \quad n > 0,$$

$$\le \frac{1}{n} \max \{0, p \log \|A^n\|\}$$

$$= \max \{0, p \log \|A^n\|^{1/n}\}.$$

But, $\|A^n\|^{1/n} \to$ the spectral radius of A, which here is precisely τ. Therefore

$$h_\rho(A) \le \max \{0, p \log \tau\}. \quad //$$

Remark:

If $A: R^p \to R^p$ is linear and the metric ρ is determined by a norm then $h_\rho(A) = \lim_{\varepsilon \to 0} \limsup_{n \to \infty} [-\frac{1}{n} \log m(D_n(0,\varepsilon,A))]$ where m is

Lebesgue measure on R^p and $D_n(0,\varepsilon,A) = \bigcap_{i=0}^{n-1} A^{-i}B_\varepsilon(0)$.

Proof: We can suppose ρ is determined by the Euclidean norm.

In the proof of (i) of Theorem 6.12 we showed

$\bar{r}_A(\varepsilon,K) \geq \lim_{n\to\infty} \sup [-\frac{1}{n} \log m(D_n(0,\varepsilon,A))]$ and hence

$h_\rho(A) \geq \lim_{\varepsilon\to 0} \lim_{n\to\infty} \sup [-\frac{1}{n} \log m(D_n(0,\varepsilon,A))]$. Let K_q be the p-cube

in R^p with center 0 and side length $2q$. If E is an (n,ε)-

separated subset of K_q then $\bigcup_{x\in E} D_n(x,\varepsilon/2,A)$ is a disjoint union

and $\bigcup_{x\in E} D_n(x,\varepsilon/2,A) = \bigcup_{x\in E} x + D_n(0,\varepsilon/2,A) \subseteq K_{q+2\varepsilon}$. Hence

$$s_n(\varepsilon/2,K_q) \cdot m(D_n(0,\varepsilon/2,A)) \leq (q+2\varepsilon)^p$$

and hence $\bar{s}_A(\varepsilon/2,K_q) \leq \lim_{n\to\infty} \sup [-\frac{1}{n} m(D_n(0,\varepsilon/2,A))]$.

Therefore $h_\rho(T) = \sup_q h_\rho(K_q) \leq \lim_{\varepsilon\to 0} \lim_{n\to\infty} \sup [-\frac{1}{n} m(D_n(0,\varepsilon/2,A))]$. //

Theorem 6.13:

Suppose $A: R^p \to R^p$ is linear and ρ is a metric coming from a

norm. Then

$$h_\rho(A) = \sum_{|\lambda_i|>1} \log |\lambda_i|$$

where $\lambda_1,\ldots,\lambda_p$ are the eigenvalues of A.

Proof: By the Jordan Decomposition Theorem (Jordan Canonical

Form), we can write R^p as a direct sum of subspaces

$$R^p = E_1 \oplus \ldots \oplus E_k$$

where $A(E_i) \subseteq E_i$ for $i = 1,\ldots,k$ and $A_i = A|_{E_i}$ has all its eigenvalues with the same norm τ_i. Thus

$$A = A_1 \oplus \ldots \oplus A_k$$

and

$$h_\rho(A) \leq \sum_{i=1}^{k} h_\rho(A_i)$$

by use of Theorem 6.8 and (since 6.8 is stated in terms of a specific metric) the fact that all norms on R^p are equivalent. By Theorem 6.12

$$h_\rho(A) \leq \sum_{i=1}^{k} \max\{0, \dim E_i \cdot \log \tau_i\}$$

$$= \sum_{\tau_i > 1} (\dim E_i \cdot \log \tau_i)$$

$$= \sum_{|\lambda_i| > 1} \log |\lambda_i|.$$

We can suppose ρ is determined by the Euclidean norm. By the above remark we have $h_\rho(A) = \lim_{\varepsilon \to 0} \limsup_{n \to \infty} [-\frac{1}{n} \log m(D_n(0,\varepsilon,A))]$. Write R^p as a direct sum of two subspaces $R^p = F_1 \oplus F_2$ so that $AF_i \subseteq F_i$ ($i = 1,2$) and $A_1 = A|_{F_1}$ has eigenvalues with absolute value greater than one and $A_2 = A|_{F_2}$ has eigenvalues with absolute value less than or equal to one. Since $D_n(0,\varepsilon,A) \subseteq B_\varepsilon(0) \cap A^{-(n-1)}B_\varepsilon(0)$ we have $m(D_n(0,\varepsilon,A)) \leq c_\varepsilon |\det A_1^{-(n-1)}|$ for some c_ε independent of n. Therefore $\limsup_{n \to \infty} [-\frac{1}{n} m(D_n(0,\varepsilon,A)] \geq \log |\det A_1| = \sum_{|\lambda_i| > 1} \log |\lambda_i|$. Therefore

$$h_\rho(A) \geq \sum_{|\lambda_i| > 1} \log |\lambda_i|. \quad //$$

Theorem 6.14:

Suppose $T: K^p \to K^p$ is an affine transformation, $Tx = a \cdot A(x)$ where $a \in K^p$ and A is a surjective endomorphism of K^p. If m is Haar measure, then

$$h(T) = h_m(T) = h_m(A) = h(A) = \sum_{|\lambda_i| > 1} \log |\lambda_i|,$$

the λ_i's being the eigenvalues of the matrix $[A]$ which represents A.

Proof: We know by Theorem 6.10 that

$$h(T) = h_m(T) = h_m(A) = h(A)$$

and by Corollary 6.11 that $h(A) = h(\tilde{A})$, where $\tilde{A}$ denotes the covering linear map of A. $h(\tilde{A})$ is calculated in Theorem 6.13. //

Note:

We have given a full proof of this result when the space is a finite-dimensional torus (since we proved Theorem 6.9 only in this case). The above proof is due to Bowen. This formula for the entropy of an automorphism was first stated by Sinai [1].

§5. Expansive Homeomorphisms

As an analogue of the measure theoretic concept of a generator, one could make the following definition:

Let (X,d) be a compact metric space, and $T: X \to X$ a homeomorphism.

Definition 6.8:

A finite open cover α of X is a generator (weak generator) for T if for every bisequence $\{A_n\}$ of members of α,

$$\bigcap_{n=-\infty}^{\infty} T^{-n} \bar{A}_n \qquad \text{is at most one point}$$

$$\left(\bigcap_{n=-\infty}^{\infty} T^{-n} A_n \qquad \text{is at most one point} \right).$$

These concepts are due to Keynes and Robertson [1].

Theorem 6.15:

    T has a generator iff T has a weak generator.

    Proof: ($\Rightarrow$) is trivial.

    ($\Leftarrow$). Let β be a weak generator for T,

$$\beta = \{B_1, \ldots, B_s\},$$

and let δ be a Lebesgue number for β. Let α be a finite open cover by sets A_i having $\text{diam}(\bar{A}_i) \leq \delta$. So if A_{i_n} is a bisequence in α then $\forall$ n $\exists$ j_n $\ni$ $\bar{A}_{i_n} \subseteq B_{j_n}$. Hence,

$$\bigcap_{-\infty}^{\infty} T^{-n} \bar{A}_{i_n} \subseteq \bigcap_{-\infty}^{\infty} T^{-n} B_{j_n}$$

which is either empty or a single point. So α is a generator. //

    The following shows that a generator determines the topology on X.

Theorem 6.16:

    Let α be a generator for T. Then $\forall$ $\varepsilon > 0$ $\exists$ $N > 0$ $\ni$ each set in $\bigvee_{-N}^{N} T^{-n} \alpha$ has diameter $< \varepsilon$. Conversely, $\forall$ $N > 0$ $\exists$ $\varepsilon > 0$ such that $d(x,y) < \varepsilon$ implies

$$x,y \in \bigcap_{-N}^{N} T^{-n} A_n$$

for some $A_{-N}, \ldots, A_N \in \alpha$.

Proof: Suppose the first part of the theorem does not hold.
$\exists\ \varepsilon > 0 \ni \forall\ j > 0\ \exists\ x_j, y_j,\ d(x_j, y_j) > \varepsilon$ and $\exists\ A_{j,i} \in \alpha,$
$-j \le i \le j$ with $x_j, y_j \in \bigcap_{i=-j}^{j} T^{-i} A_{j,i}.$ We can suppose that $x_j \to x,$
$y_j \to y$ since X is compact, and hence $x \ne y.$ Consider the sets
$A_{j,0}.$ Infinitely many of them coincide since α is finite. Thus
$x_j, y_j \in A_0,$ say, for infinitely many j and hence $x, y \in \bar{A}_0.$
Similarly, for each $n,$ infinitely many $A_{j,n}$ coincide and we obtain $A_n \in \alpha$ with $x, y \in T^{-n}\bar{A}_n.$ Thus,

$$x, y \in \bigcap_{-\infty}^{\infty} T^{-n}\bar{A}_n$$

contradicting the fact that α is a generator.

To prove the converse let $N > 0$ be given. Let $\delta > 0$ be a
Lebesgue number for $\alpha.$ Choose $\varepsilon > 0$ such that $d(x,y) < \varepsilon$ implies
$d(T^i x, T^i y) < \delta$ for $-N \le i \le N.$ Hence if $d(x,y) < \varepsilon$ and $|i| \le N$
then $T^i x, T^i y \in A_i$ for some $A_i \in \alpha.$ Hence

$$x, y \in \bigcap_{-N}^{N} T^{-i} A_i. \quad //$$

The analogue of the Kolmogorov-Sinai Theorem is:

Theorem 6.17:

If α is a generator for T then

$$h(T) = h(T, \alpha).$$

Proof: Let β be any open cover. Let δ be a Lebesgue number
for $\beta.$ Choose $N > 0$ so that each member of $\bigvee_{-N}^{N} T^{-n}\alpha$ has diameter
$< \delta.$ Then $\beta < \bigvee_{-N}^{N} T^{-n}\alpha,$ and so,

$$h(T,\beta) \leq h(T, \overset{N}{\underset{-N}{\bigvee}} T^{-n}\alpha)$$

$$= \lim_{k\to\infty} \frac{1}{k} H(\overset{n-1}{\underset{i=0}{\bigvee}} T^{-i}(\overset{N}{\underset{-N}{\bigvee}} T^{-n}\alpha))$$

$$= \lim_{k\to\infty} \frac{1}{k} H(\overset{N+k-1}{\underset{-N}{\bigvee}} T^{-n}\alpha)$$

$$= \lim_{k\to\infty} \frac{1}{k} H(\overset{2N+k-1}{\underset{0}{\bigvee}} T^{-n}\alpha)$$

$$= \lim_{k\to\infty} \frac{2N+k-1}{k} \cdot \frac{1}{2N+k-1} H(\overset{2N+k-1}{\underset{0}{\bigvee}} T^{-n}\alpha)$$

$$= h(T,\alpha).$$

Therefore, $h(T,\beta) \leq h(T,\alpha)$ for all open covers β. Hence

$$h(T) = h(T,\alpha). \quad //$$

Remark:

The same result holds for weak generators.

Generators are connected with the notion of expansive homeomor-phism, which was studied long ago.

Definition 6.10:

A homeomorphism $T: X \to X$ is <u>expansive</u> if $\exists$ $\delta > 0$ $\ni$ if $x \neq y$ then $\exists$ $n \in Z$ $\ni$ $d(T^n x, T^n y) \geq \delta$. We call δ an <u>expansive constant</u> for T.

Remark:

Another way to define an expansive homeomorphism is as follows. Consider $X \times X$ with $T \times T$ acting on it. Define a metric D on $X \times X$ by $D((u,v),(x,y)) = \max \{d(u,x),d(v,y)\}$. Then T is expansive $\leftrightarrow$ $\exists$ $\delta > 0$ such that if (x,y) is not an element of the diagonal, some power of T×T takes (x,y) out of the δ-neighborhood of the diagonal.

The following theorem is due to Reddy, and Keynes and Robertson.

Theorem 6.18:

T is expansive iff T has a generator iff T has a weak generator.

Proof: By Theorem 6.15 it suffices to show T is expansive iff T has a generator.

(⇒) Let δ be an expansive constant for T and α a finite cover by open balls of radius δ/2. Suppose x,y ∈ $\bigcap_{-\infty}^{\infty} T^{-n}\bar{A}_n$ where $A_n ∈ α$. Then, $d(T^n x, T^n y) ≤ δ$ ∀ n, so, by assumption x = y. Therefore α is a generator.

(⇐) Conversely, suppose α is a generator. Let δ be a Lebesgue number for α. If $d(T^n x, T^n y) ≤ δ$ ∀ n then ∀ n ∃ $A_n ∈ α$ ∋ $T^n x, T^n y ∈ A_n$ and so,

$$x,y ∈ \bigcap_{-\infty}^{\infty} T^{-n} A_n$$

which is at most one point. Hence x = y and T is expansive. //

Corollary 6.18:

(1) Expansiveness is independent of the metric (however, the expansive constant does change).

(2) T is expansive iff T^k is expansive, k ≠ 0.

(3) Expansiveness is a topological conjugacy invariant.

Proof: (1) This is trivial, since having a generator has nothing to do with the metric.

(2). If α is a generator for T then

$$α ∨ T^{-1}α ∨ ... ∨ T^{-(k-1)}α$$

is a generator for T^k. If α is a generator for T^k then α is also a generator for T.

(3) is trivial. //

The next result shows how to find measure theoretic generators for expansive homeomorphisms.

<u>Theorem 6.19:</u>

Let T be expansive with constant δ. If

$$\xi = \{C_1,\ldots,C_s\}$$

is a partition of X into Borel sets of diameter $< \delta$, then ξ is a measure theoretic generator for any T-invariant Borel probability measure.

<u>Proof</u>: Let C_{i_n} be a bisequence of members of ξ. If $x,y \in \bigcap_{-\infty}^{\infty} T^{-n}C_{i_n}$ then $T^n x, T^n y \in C_{i_n}$ for all n, and hence $d(T^n x, T^n y) < \delta$ $\forall$ n. By expansiveness $x = y$. Thus $\bigcap_{-\infty}^{\infty} T^{-n}C_{i_n} = \phi$ or = one point. Hence

$$\bigvee_{-\infty}^{\infty} T^n A(\xi) = B. \quad //$$

Thus, for expansive homeomorphisms there are many measure theoretic generators.

<u>Examples:</u>

(1) Isometries are never expansive except on finite spaces.

(2) Let A be an automorphism of the n-torus, and [A] the corresponding matrix. Then A is expansive iff [A] has no eigenvalues of modulus 1.

<u>Sketch of proof</u>: One first shows that A is expansive iff the linear map $\tilde{A}$ of R^n that covers A is expansive. Then show that

$\tilde{A}$ is expansive iff the complexification of $\tilde{A}$ is expansive. Then one shows that the complexification of $\tilde{A}$ is expansive iff the transformation given by the Jordan normal form is expansive. Lastly, one shows that the normal form is expansive iff there are no eigenvalues of modulus 1.

(Note: By Theorem 6.19, any partition of K^n into sufficiently small n-rectangles is a measure theoretic generator for an expansive automorphism of K^n.)

(3) The two-sided shift on k symbols is expansive.

Proof (1): Let the state space be $\{0,1,\ldots,k-1\}$. Let $A_i = \left\{\{x_n\}: x_0 = i\right\}$, $i = 0,1,\ldots,k-1$. Then $A_0 \cup A_1 \cup \ldots \cup A_{k-1} = X$ and each A_i is open. $\alpha = \{A_0,\ldots,A_{k-1}\}$ is a generator for the shift since if $x \in \bigcap_{-\infty}^{\infty} T^{-n} A_{i_n}$ where the $A_{i_n} \in \alpha$ then

$$x = (\ldots,i_{-2},i_{-1},i_0,i_1,i_2,\ldots).$$

We then use Theorem 6.18. //

Proof (2): Let d be the metric given by:

$$d(\{x_n\},\{y_n\}) = \sum_{n=-\infty}^{\infty} \frac{|x_n - y_n|}{2^{|n|}}.$$

Suppose $\{x_n\} \neq \{y_n\}$. Then for some n_0, $x_{n_0} \neq y_{n_0}$ and

$$d(T^{n_0}\{x_n\}, T^{n_0}\{y_n\}) = \sum_{n=-\infty}^{\infty} \frac{1}{2^{|n|}} |x_{n+n_0} - y_{n+n_0}|$$

$$\geq |x_{n_0} - y_{n_0}| \geq 1.$$

Thus 1 is an expansive constant. //

Remarks:

(a) If $T: X \to X$ is expansive and Y is a closed subset of X with

$TY = Y$ then $T|_Y$ is expansive.

(b) If $T_1: X_1 \rightarrow X_1$, $T_2: X_2 \rightarrow X_2$ are expansive then so is $T_1 \times T_2: X_1 \times X_2 \rightarrow X_1 \times X_2$. Any finite direct product of expansive homeomorphisms is expansive, but infinite products are not.

(c) If $T: X \rightarrow X$ is expansive and $S: Y \rightarrow Y$ is a homeomorphism with

$$
\begin{array}{ccc}
 & T & \\
X & \longrightarrow & X \\
\phi \downarrow & & \downarrow \phi \\
Y & \longrightarrow & Y \\
 & S &
\end{array}
$$

commutative for $\phi: X \rightarrow Y$ a continuous surjection then S need not be expansive. S is expansive if ϕ is a k-to-one covering map. So expansiveness is not preserved under the operation of taking factors.

Example: (Parry and Walters)

Consider the 2-torus K^2; identify (z,w) with $(\bar{z},\bar{w})$. The map ϕ is a 2-to-one covering map except at four points:

$$(1,1), \ (-1,1), \ (-1,-1), \ \text{and} \ (1,-1).$$

Let A: $K^2 \rightarrow K^2$ be an automorphism. The quotient space is S^2 and A induces a homeomorphism on S^2 which can be shown to be non-expansive.

Another example can be constructed as follows:

Let $Tz = az$ be a minimal rotation of K. We shall represent T as a factor of a subset of the two-sided shift on two symbols. Consider the cover of K by the closed intervals (arcs) between -1 and 1 on K. Call one of them A_0 and the other A_1. If $z \in K \setminus \{a^n, -a^n: n \in Z\}$ we can uniquely associate a member of $\prod_{-\infty}^{\infty} \{0,1\}$ to z by $z \rightarrow \{a_n\}_{-\infty}^{\infty}$ if $T^n z \in A_{a_n}$. Let Λ denote the subset of $\prod_{-\infty}^{\infty} \{0,1\}$

arising in this way. The map

$$\phi: \Lambda \to K \setminus \{a^n, -a^n: n \in Z\}$$

defined above satisfies $\phi S(x) = T\phi(x)$, $x \in \Lambda$ where S denotes the shift, and if we can show ϕ is uniformly continuous then ϕ extends uniquely to a continuous map $\pi: \bar{\Lambda} \to K$ with $\pi S = T\pi$. Suppose $\varepsilon > 0$ is given. Choose $N > 0$ so that $\{1, a^{\pm 1}, a^{\pm 2}, \ldots, a^{\pm N}\}$ is $\varepsilon/2$-dense in K. Suppose $\{b_n\}$ and $\{c_n\}$ are two members of Λ such that $b_n = c_n$ for $|n| \le N$. We then have to show $d(\phi(\{b_n\}), \phi(\{c_n\})) < \varepsilon$. Let $x = \phi(\{b_n\})$ and $y = \phi(\{c_n\})$. The assumption $b_n = c_n$, $|n| \le N$ means that $a^n x$ and $a^n y$ belong to the same element of the cover for $|n| \le N$. If $y = -x$ then this clearly cannot happen. So suppose the counter-clockwise distance from y to x is smaller than the clockwise distance. For some n with $|n| \le N$ $a^n x$ is in the open interval of length ε starting at 1 and going counter-clockwise. Hence $a^n y$ must also be in the upper half of the circle and by the assumption about the relative positions of x and y, $a^n y$ must be between 1 and $a^n x$. Hence $d(a^n x, a^n y) < \varepsilon$ and so $d(x, y) < \varepsilon$.

We shall now show that every expansive homeomorphism is a factor of a subset of a two-sided shift.

Theorem 6.20:

Let $T: X \to X$ be an expansive homeomorphism. Then $\exists$ an integer $k > 0$, a closed subset Ω of

$$X_k = \prod_{-\infty}^{\infty} \{0, 1, \ldots, k-1\}$$

such that $\sigma\Omega = \Omega$, where σ is the shift on X_k, and a continuous surjection $\pi: \Omega \to X$ such that

$$\pi\sigma(y) = T\pi(y) \qquad y \in \Omega.$$

Proof: The proof will resemble that of the preceding example.
Let δ be an expansive constant for T. Choose a cover $\alpha = \{A_0,\ldots,A_{k-1}\}$ by closed sets such that $\text{diam}(A_i) < \delta$ $\forall$ i and so that the A_i intersect only in their boundaries. Let D denote the union of the boundaries of the A_i. Then $D_\infty = \bigcup_{-\infty}^{\infty} T^n D$ is a first category set and so $X \backslash D_\infty$ is dense in X. For each $x \in X \backslash D_\infty$ we can assign, uniquely, a member of X_k by $x \to \{a_n\}_{-\infty}^{\infty}$ iff $T^n x \in A_{a_n}$.
Let Λ denote the collection of all sequences arising in this way.
If $\phi: \Lambda \to X \backslash D_\infty$ is the map defined above then $\phi\sigma(y) = T\phi(y)$ $\forall$ $y \in \Lambda$ and if we can show ϕ is uniformly continuous it will then follow that ϕ can be uniquely extended to a continuous map $\pi: \bar{\Lambda} \to X$ such that $\pi\sigma(y) = T\pi(y)$ $\forall$ $y \in \bar{\Lambda}$.

Let $\varepsilon > 0$ be given. Choose $N > 0$ so that each member of $\bigvee_{-N}^{N} T^n \alpha$ has diameter less than ε, which can be done by Theorem 6.16, since we can enlarge each A_i to an open set to obtain a generator (remembering $\text{diam}(A_i) < \delta$). If $\{a_n\},\{b_n\} \in \Lambda$ and $a_n = b_n$ for $|n| \le N$ then $\phi(\{a_n\})$, $(\{b_n\})$ are in the same member of $\bigvee_{-N}^{N} T^n \alpha$ and so $d(\phi\{a_n\},\phi\{b_n\}) < \varepsilon$. Hence ϕ is uniformly continuous. //

The periodic points of T are associated with a generator as follows:

Theorem 6.21:
Let $T: X \to X$ be expansive and let α be a generator (or a weak generator). Then $T^k x = x$ iff

$$x = \bigcap_{i=-\infty}^{\infty} T^{k \cdot i}(A_0 \cap TA_1 \cap \ldots \cap T^{k-1}A_{k-1})$$

where $A_j \in \alpha$ $j = 0,\ldots,k-1$.

Proof: Suppose $T^k x = x$. Since $\alpha \vee T\alpha \vee \ldots \vee T^{k-1}\alpha$ is a cover of X, $x \in A_0 \cap TA_1 \cap \ldots \cap T^{k-1}A_{k-1}$ for some collection of A_j's in α. Thus

$$x \in T^{k \cdot i}(A_0 \cap TA_1 \cap \ldots \cap T^{k-1}A_{k-1}) \quad \forall \ i,$$

and this implies

$$x \in \bigcap_{-\infty}^{\infty} T^{k \cdot i}(A_0 \cap TA_1 \cap \ldots \cap T^{k-1}A_{k-1})$$

which is at most one point. Therefore,

$$x = \bigcap_{-\infty}^{\infty} T^{k \cdot i}(A_0 \cap TA_1 \cap \ldots \cap T^{k-1}A_{k-1}).$$

The converse is trivial. //

The following gives an estimate on the number of fixed points of T^n.

Corollary 6.21:

If T is expansive and α is a generator for T with M members, then

$$N_n(T) = |\{x : T^n x = x\}| \leq M^n \qquad (n > 0).$$

Topological entropy is connected to periodic points by

Theorem 6.22:

If T is expansive, then

$$h(T) \geq \omega(T) = \overline{\lim_{n \to \infty}} \frac{1}{n} \log N_n(T).$$

Proof: Let α be a generator for T. Each element of $\bigvee_0^{n-1} T^{-i}\alpha$ contains at most one point fixed by T^n, since if $T^n x = x$, $T^n y = y$

and $x,y \in \bigcap_{i=0}^{n-1} T^{-i}A_{j_i}$ then

$$x,y \in \bigcap_{k=-\infty}^{\infty} T^{-nk}(A_{j_0} \cap T^{-1}A_{j_1} \cap \ldots \cap T^{-(n-1)}A_{j_{n-1}})$$

which is at most one point.   Therefore $x = y$.   Thus,

$$N_n(T) \leq N(\alpha \vee T^{-1}\alpha \vee \ldots \vee T^{-(n-1)}\alpha)$$

which implies that

$$\frac{1}{n} \log N_n(T) \leq \frac{1}{n} H(\alpha \vee T^{-1}\alpha \vee \ldots \vee T^{-(n-1)}\alpha)$$

so,

$$\overline{\lim_{n \to \infty}} \frac{1}{n} \log N_n(T) \leq h(T,\alpha) = h(T).$$

Therefore $\qquad\qquad\qquad \omega(T) \leq h(T).$ //

Consider our examples:

(1)   Let A be an expansive automorphism of the torus K^m.

$$N_n(A) = |\{x: A^n x = x\}| = |\text{Kernel } (A^n - I)|$$

$$= |\det ([A]^n - I)| \quad \text{(by the proposition below)}$$

$$= |\prod_i (\lambda_i^n - 1)|$$

where the λ_i are the eigenvalues of the matrix $[A]$.   So,

$$\omega(A) = \lim_{n \to \infty} \frac{1}{n} \sum_{\lambda_i} (\log |\lambda_i^n - 1|).$$

If $|\lambda_i| > 1$ then

$$\frac{1}{n} \log |\lambda_i^n - 1| = \frac{1}{n} [\log |\lambda_i|^n + \log |1 - \lambda_i^{-n}|]$$

$$\to \log |\lambda_i| + 0 = \log |\lambda_i|.$$

If $|\lambda_i| < 1$ then

$$\frac{1}{n} \log |\lambda_i^n - 1| \to 0.$$

So, $\qquad \omega(A) = \sum_{|\lambda_i|>1} \log |\lambda_i| = h(A) \qquad$ by Theorem 6.14.

Therefore, in this case we have equality.

(2) If T is the two-sided shift on k symbols then

$$N_n(T) = k^n \quad \text{and} \quad \omega(T) = \log k = h(T),$$

so that here too we have equality.

In these examples $\omega(T)$ and $h(T)$ coincide. However, this is not true for all expansive homeomorphisms as there are examples of minimal expansive homeomorphisms ($\omega(T) = 0$ since minimality implies there are no periodic points when the space is infinite) with positive topological entropy (due to Furstenberg).

Problem:

If T is expansive and has a dense set of periodic points then is it true that $\omega(T) = h(T)$?

Bowen has shown this to be true under the stronger assumption that T is an axiom A^* homeomorphism.

In example (1) we used the following:

Proposition:

If $B: K^n \to K^n$ is an endomorphism of K^n onto K^n with corresponding matrix $[B]$, (so $\det [B] \neq 0$) then B is a $|\det [B]|$-to-one map.

(We used the case $B = A^n - I$ where $[A]$ has no roots of unity as eigenvalues.)

Proof: [B] is an invertible matrix since det [B] ≠ 0. Thus, we can write [B] = $E_1 E_2 \ldots E_n$ where the E_i are elementary matrices. Since [B] has integer entries, the E_i have rational entries. Each E_i is one of the following forms:

(1) I with two rows interchanged,

(2) I with one row multiplied by $c \in Q$, or

(3) I with the j-th row replaced by j-th row + c(k-th row), $c \in Q$.

For each E_i, choose $e_i \in Z$ such that $e_i E_i$ has integer entries. In case (1), $e_i E_i$ induces an $|e_i|^n$-to-one map of K^n, i.e., an $|e_i|^n |\det E_i|$-to-one map. In case (2), $e_i E_i$ induces an $|e_i|^n |c|$-to-one map of K^n, i.e., an $|e_i|^n |\det E_i|$-to-one map. In case (3), $e_i E_i$ induces an $|e_i|^n |\det E_i|$-to-one map. Let C_i be the endomorphism of K^n determined by $e_i E_i$. Let B' be the endomorphism determined by $e_1 \ldots e_r [B]$. Then $B' = C_1 \circ \ldots \circ C_r$. If B' is a b'-to-one map and B is a b-to-one map then since $B' = C_1 \circ \ldots \circ C_r$

$$b' = \prod_{i=1}^{r} |e_i|^n |\det E_i|$$

$$= \prod_{i=1}^{r} |e_i|^n \cdot \prod_{i=1}^{r} |\det E_i|$$

$$= \prod_{i=1}^{r} |e_i|^n |\det [B]|.$$

But, also

$$b' = \prod_{i=1}^{r} |e_i|^n \cdot b$$

since $[B'] = e_1 \ldots e_r [B]$, so

$$b = |\det [B]|. \quad //$$

Consider $\quad \omega(T) = \overline{\lim_{n \to \infty}} \frac{1}{n} \log N_n(T)$.

$\omega(T)$ is connected with the ζ-function which was introduced (for diffeomorphisms) by Artin and Mazur [1]:

If $T: X \to X$ is a homeomorphism such that $N_n(T) < \infty$ for all $n > 0$ then we set

$$\zeta_T(z) = \exp\left(\sum_{n=1}^{\infty} \frac{1}{n} z^n N_n(T)\right), \quad z \in C.$$

Note that

$$\omega(T) = \log\left(\frac{1}{\text{radius of convergence of } \zeta_T}\right).$$

(Artin and Mazur showed that "most" diffeomorphisms have a ζ-function with positive radius of convergence, and Smale suggested that the ζ-function might be a rational function of z for "most" diffeomorphisms. Manning [1] has shown this to be the case for axiom A diffeomorphisms but results of Simon [1] have answered Smale's conjecture negatively (for K^3).)

It is known that there are no expansive homeomorphisms of S^1, but not known if there are any on S^2. It seems reasonable to ask whether a compact metric space admitting an expansive homeomorphism is finite-dimensional.

§6. Examples

We consider the topological entropy of some examples.

(1) Isometries have zero entropy. This is clear from Bowen's definition. Hence rotations on compact metric groups and all topologically transitive homeomorphisms with topological discrete spectrum have zero entropy (Theorem 5.8).

(2) The two-sided shift on k symbols has entropy $\log k$. This is proved by considering the obvious generator.

(3) Any homeomorphism of K has zero topological entropy.

Proof: Let $T: K \to K$ be a homeomorphism. T maps intervals to intervals as the intervals are the connected subsets. Suppose the circle has length 1.

Choose $\varepsilon > 0$ such that

$$d(x,y) \le \varepsilon \;\Rightarrow\; d(T^{-1}x, T^{-1}y) \le 1/4.$$

Consider spanning sets for K with respect to T. Clearly, $r_1(\varepsilon, K) \le [1/\varepsilon] + 1$, where $[\cdot]$ denotes the least integer function. We estimate $r_n(\varepsilon, K)$.

Suppose we have an $(n-1, \varepsilon)$-spanning set F of minimal cardinality $r_{n-1}(\varepsilon, K)$. Consider the points of $T^{n-1}F$ and the intervals they determine. Add points to this set so that the new intervals have length $< \varepsilon$. We have added at most $\left[\frac{1}{\varepsilon}\right] + 1$ points. Let

$$F' = F \cup T^{-(n-1)}(\text{these new points}).$$

We claim that F' is an (n, ε)-spanning set for K. Let $x \in K$. Then $\exists\; y \in F \ni$

$$\max_{0 \le i \le n-2} d(T^i x, T^i y) \le \varepsilon.$$

If $d(T^{n-1}x, T^{n-1}y) \le \varepsilon$ then our claim is proved. If there is no $y \in F$ with both these properties, choose a $y \in F \ni$

$$\max_{0 \le i \le n-2} d(T^i x, T^i y) \le \varepsilon.$$

Consider the interval between $T^{n-1}x$ and $T^{n-1}y$ which is mapped by T^{-1} to the ε-interval $[T^{n-2}x, T^{n-2}y]$. Choose a point $T^{n-1}z$, $z \in F'$ inside the chosen interval $[T^{n-1}x, T^{n-1}y]$ and with $d(T^{n-1}x, T^{n-1}z) \le \varepsilon$. Then $T^{n-2}z$ lies in the ε-interval $[T^{n-2}x, T^{n-2}y]$ and so, $d(T^{n-2}z, T^{n-2}x) \le \varepsilon$. The ε-interval $[T^{n-2}x, T^{n-2}y]$ is mapped by T^{-1} to an interval of length $\le 1/4$, and hence to the ε-interval

$[T^{n-3}x, T^{n-3}y]$. So, since $T^{n-3}z$ is in this interval, $d(T^{n-3}x, T^{n-3}z) \le \varepsilon$. Similarly, by induction

$$d(T^i x, T^i z) \le \varepsilon \quad \forall \ i, \quad 0 \le i \le n-1.$$

Thus, F' is an (n, ε)-spanning set for K. So,

$$r_n(\varepsilon, K) \le r_{n-1}(\varepsilon, K) + [1/\varepsilon] + 1$$

$$\le n([1/\varepsilon] + 1).$$

Therefore, $\qquad \bar{r}_T(\varepsilon, K) = \overline{\lim_{n \to \infty}} \frac{1}{n} \log r_n(\varepsilon, K) = 0,$

so, $\qquad\qquad\qquad h(T) = 0. \quad //$

Corollary:

Any homeomorphism of $[0,1]$ has zero topological entropy.

Proof: $T: [0,1] \to [0,1]$ has either $T(0) = 0$ and $T(1) = 1$ or $T(0) = 1$ and $T(1) = 0$. In both cases T^2 induces a homeomorphism of K. $//$

(4) If $T: M^m \to M^m$ is a differentiable map of an m-dimensional Riemannian manifold M^m with Riemannian metric $\|\cdot\|$, then

$$h_\rho(T) \le \max \{0, m \log \sup_{x \in M} \|dT_x\|\}$$

where $dT_x: M_x \to M_{T(x)}$ is the derivative of T at x and ρ is the metric on M determined by the Riemannian metric. This has been proved by several people.

$$--\Omega--$$

Bibliography

R. Abraham:

[1] <u>Foundations of Mechanics</u>, Benjamin, 1967.

R. Abraham and J. Robbin:

[1] <u>Mappings and Flows</u>, Benjamin, 1967.

R. L. Adler:

[1] Skew products of Bernoulli shifts and rotations, Israel
     Journal of Mathematics, vol. 12, pp. 215-222, 1972.

R. L. Adler, A. G. Konheim and M. H. McAndrew:

[1] Topological entropy, Transactions of the American Mathema-
     tical Society, vol. 114, pp. 309-319, 1965.

R. L. Adler and P. C. Shields:

[1] Skew products of Bernoulli shifts with rotations, Israel
     Journal of Mathematics, vol. 12, pp. 215-222, 1972.

R. L. Adler and B. Weiss:

[1] <u>Similarity of Automorphisms of the Torus</u>, Memoirs of the
     American Mathematical Society, no. 98, 1970.

H. Anzai:

[1] Ergodic skew-product transformations on the torus, Osaka
     Mathematics Journal, vol. 3, pp. 83-99, 1951.

M. Artin and B. Mazur:

[1] On periodic points, Annals of Mathematics, vol. 81, pp. 82-
     99, 1965.

L. Auslander, L. Green, and F. Hahn:

[1] <u>Flows on Homogeneous Spaces</u>, Annals of Mathematics Studies,
     no. 53, 1963.

A. Avez and V. I. Arnold:

[1] Ergodic Problems in Classical Mechanics, Benjamin, 1968.

K. Berg:

[1] Convolution of invariant measures, maximal entropy,
     Mathematical Systems Theory, vol. 3, pp. 146-150, 1969.

P. Billingsley:

[1] Ergodic Theory and Information, Wiley, 1965.

G. D. Birkhoff:

[1] Proof of the ergodic theorem, Proceedings of the National
     Academy of Sciences USA, vol. 17, pp. 656-660, 1931.

J. R. Blum and D. L. Hanson:

[1] On the isomorphism problem for Bernoulli schemes, Bulletin
     of the American Mathematical Society, vol. 69, pp. 221-223,
     1963.

R. Bowen:

[1] Topological entropy and axiom A, Proceedings of the Summer
     Institute on Global Analysis, Berkeley, California, pp. 23-
     42, 1968.

[2] Markov partitions for axiom A diffeomorphisms, American
     Journal of Mathematics, vol. 92, pp. 725-747, 1970.

[3] Markov partitions and minimal sets for axiom A diffeomor-
     phisms, American Journal of Mathematics, vol. 92, pp. 907-
     918, 1970.

[4] Entropy for group endomorphisms and homogeneous spaces,
     Transactions of the American Mathematical Society, vol. 153,
     pp. 401-414, 1971.

[5] Periodic points, measures and axiom A, Transactions of the
     American Mathematical Society, vol. 154, pp. 377-397, 1971.

[6] Periodic orbits for hyperbolic flows, American Journal of
     Mathematics, vol. 94, pp. 1-30, 1972.

[7] Entropy-expansive maps, Transactions of the American Mathe-
     matical Society, vol. 164, pp. 323-331, 1972.

[8] The equidistribution of closed geodesics, American Journal
     of Mathematics, vol. 94, pp. 413-423, 1972.

[9] One-dimensional hyperbolic sets for flows, Journal of Dif-
     ferential Equations, vol. 12, pp. 173-179, 1972.

[10] Symbolic dynamics for hyperbolic flows, American Journal
of Mathematics, vol. 45, pp. 429-460, 1973.

[11] Topological entropy for noncompact sets, Transactions of
the American Mathematical Society, vol. 184, pp. 125-136,
1973.

[12] Entropy versus homology for certain diffeomorphisms,
Topology, vol. 13, pp. 61-67, 1974.

R. Bowen and P. Walters:

[1] Expansive one-parameter flows, Journal of Differential
Equations, vol. 12, pp. 180-193, 1972.

R. V. Chacon:

[1] Change of velocity in flows, Journal of Mathematics and
Mechanics, vol. 16, pp. 417-431, 1966.

Hsin Chu:

[1] Some results on affine transformations of compact groups,
to appear.

J. P. Conze:

[1] Entropie des flots et des transformations affine sur les
espaces homogenes, Compte Rendu, vol. 270, pp. 547-548,
1970.

E. I. Dinaberg:

[1] The relation between topological entropy and metric entropy,
Doklady Akademii Nauk SSSR, vol. 190, 1970 (Russian),
Soviet Mathematics, vol. 11, pp. 13-16, 1970 (English).

[2] A connection between various entropy characterizations of
dynamical systems, Izvestija Akademii Nauk SSSR, Serija
Matematiceskaja, vol. 35, pp. 324-366, 1971 (Russian).

R. Ellis:

[1] Lectures on Topological Dynamics, Benjamin, 1969.

N. A. Friedman:

[1] Introduction Ergodic Theory, Van Nostrand, 1970.

[2] Bernoulli shifts induce Bernoulli shifts, Advances in
Mathematics, vol. 10, pp. 29-48, 1973.

N. A. Friedman and D. S. Ornstein:

[1] An isomorphism of weak Bernoulli Transformations, Advances
     in Mathematics, vol. 5, pp. 365-394, 1970.

[2] Entropy and the Isomorphism Problem, to be published by
     Springer.

[3] Ergodic transformations induce mixing transformations,
     Advances in Mathematics, vol. 10, pp. 147-163, 1973.

H. Furstenberg:

[1] Strict ergodicity and transformations of the torus,
     American Journal of Mathematics, vol. 83, pp. 573-601, 1961.

[2] Disjointness in ergodic theory, Mathematical Systems Theory,
     vol. 1, pp. 1-50, 1967.

T. N. T. Goodman:

[1] Relating topological entropy and measure entropy, Bulletin
     of the London Mathematical Society, vol. 3, pp. 176-180.
     1971.

[2] Maximal measures for expansive homeomorphisms, Journal of
     the London Mathematical Society, vol. 5, pp. 439-444, 1972.

L. W. Goodwyn:

[1] Topological entropy bounds measure-theoretic entropy,
     Proceedings of the American Mathematical Society, vol. 23,
     pp. 679-688, 1969.

[2] A characterization of symbolic cascades in terms of expan-
     siveness and topological entropy, Mathematical Systems
     Theory, vol. 4, pp. 157-157, 1970.

[3] Some counterexamples in topological entropy, Topology,
     vol. 11, pp. 377-385, 1972.

[4] Comparing topological entropy with measure theoretic
     entropy, American Journal of Mathematics, vol. 74, pp. 366-
     388, 1972.

W. H. Gottschalk:

[1] Bibliography for Dynamical Topology, fourth edition,
     Wesleyan University, 1969.

W. H. Gottschalk and G. A. Hedlund:

[1] Topological Dynamics, American Mathematical Society Col-
     loquium Publications, 1955.

C. Grillenberger:

[1]   Constructions of strictly ergodic systems I, II, Zeit-
      schrift für Wahrscheinlichkeitstheorie, vol. 25, pp. 323-
      334, 1973.

F. Hahn:

[1]   On affine transformations of compact abelian groups, Ameri-
      can Journal of Mathematics, vol. 85, pp. 428-446, 1963.

F. Hahn and Y. Katznelson:

[1]   On the entropy of uniquely ergodic transformations, Trans-
      actions of the American Mathematical Society, vol. 126,
      pp. 335-360, 1967.

F. Hahn and W. Parry:

[1]   Minimal dynamical systems with quasi-discrete spectrum,
      Journal of the London Mathematical Society, vol. 40,
      pp. 309-323, 1965.

[2]   Some characteristic properties of dynamical systems with
      quasi-discrete spectrum, Mathematical Systems Theory,
      vol. 2, pp. 179-190, 1968.

P. R. Halmos:

[1]   Measure Theory, Van Nostrand, 1950.

[2]   Lectures on Ergodic Theory, Chelsea, 1953.

[3]   Introduction to Hilbert Space and the Theory of Spectral
      Multiplicity, Chelsea, 1957.

[4]   On automorphisms of compact groups, Bulletin of the Ameri-
      can Mathematical Society, vol. 49, pp. 619-624, 1943.

P. R. Halmos and J. Von Neumann:

[1]   Operator methods in classical mechanics, II, Annals of
      Mathematics, vol. 43, pp. 235-247, 1942.

H. Hoare and W. Parry:

[1]   Affine transformations with quasi-discrete spectrum, I,
      Journal of the London Mathematical Society, vol. 41,
      pp. 88-96, 1966.

[2] Affine transformations with quasi-discrete spectrum, II, Journal of the London Mathematical Society, vol. 41, pp. 529-530, 1966.

[3] Semi-groups of affine transformations, Quarterly Journal of Mathematics, Oxford, vol. 17, pp. 106-111, 1966.

E. Hopf:

[1] Ergodentheorie, Chelsea, 1937.

K. Jacobs:

[1] Neue Methode und Ergebnisse der Ergodentheorie, Springer, 1960.

[2] Lecture Notes on Ergodic Theory, Aarhus University, 1962-3.

R. Jewett:

[1] The prevalence of uniquely ergodic systems, Journal of Mathematics and Mechanics, pp. 717-729, 1970.

S. A. Juzvinskii:

[1] Calculation of the entropy of a group endomorphism, Sibirskii Matematiceskii Zurnal, vol. 8, pp. 230-239, 1967 (Russian); Siberian Mathematical Journal, vol. 8, pp. 172-178, 1967 (English).

[2] Metric properties of endomorphisms of compact groups, Izvestija Akademii Nauk SSSR, Serija Matemasiceskaja, vol. 29, pp. 1295-1328, 1965 (Russian); American Mathematical Society Translations, Series 2, vol. 66, pp. 63-98, 1968 (English).

S. Kakutani:

[1] Induced measure preserving transformations, Proceedings of the Imperial Academy of Tokyo, vol. 19, pp. 635-641, 1943.

[2] Examples of ergodic measure preserving transformations which are weakly mixing but not strongly mixing, Recent Advances in Topological Dynamics, Springer Lecture Notes, no. 318, pp. 143-149, 1973.

Y. Katznelson:

[1] Ergodic automorphisms of T^n are Bernoulli shifts, Israel Journal of Mathematics, vol. 10, pp. 186-195, 1971.

Y. Katznelson and B. Weiss:

[1] Commuting measure-preserving transformations, Israel Jour-
     nal of Mathematics, vol. 12, pp. 161-173, 1972.

[2] Ergodic automorphisms of the solenoid are Bernoulli,
     unpublished.

J. L. Kelley:

[1] General Topology, Van Nostrand, 1955.

H. B. Keynes and J. B. Robertson:

[1] Generators for topological entropy and expansiveness,
     Mathematical Systems Theory, vol. 3, pp. 51-59, 1969.

A. I. Khinchin:

[1] Mathematical Foundations of Statistical Mechanics, Dover,
     1949.

A. N. Kolmogorov:

[1] A new metric invariant of transient dynamical systems and
     automorphisms of Lebesgue spaces, Doklady Akademii Nauk
     SSSR, vol. 119, pp. 861-864, 1958.

[2] On the entropy per time unit as a metric invariant of
     automorphisms, Doklady Akademii Nauk SSSR, vol. 124,
     pp. 754-755, 1959.

W. Krieger:

[1] On entropy and generators of measure-preserving transfor-
     mations, Transactions of the American Mathematical Society,
     vol. 149, pp. 453-464, 1970.

[2] On unique ergodicity, Proceedings of the Sixth Berkeley
     Symposium on Mathematical Statistics and Probability, 1970.

N. Kryloff and N. Bogoliouboff:

[1] La théorie générale de la mesure dans son application à
     l'étude des systemes dynamique de la mécanique non linéaire,
     Annals of Mathematics, vol. 38, pp. 65-113, 1937.

A. G. Kushnirenko:

[1] Metric invariants of entropy type, Uspehi Matematiceskih
     Nauk, vol. 22, no. 5, pp. 57-65, 1967 (Russian), Russian
     Mathematical Surveys, vol. 22, no. 5, pp. 53-61, 1967
     (English).

D. A. Lind:

[1] Ergodic automorphisms of the infinite torus are Bernoulli, to appear.

G. W. Mackey:

[1] Ergodic Theory and its significance for statistical mechanics and probability theory, Advances in Mathematics, vol. 12, pp. 178-268, 1974.

A. Manning:

[1] Axiom A diffeomorphisms have rational zeta functions, Bulletin of the London Mathematical Society, vol. 3, pp. 215-220, 1971.

L. Markus:

[1] Lectures in Differentiable Dynamics, American Mathematical Society Regional Conference Series, no. 3, 1971.

R. McCabe and P. C. Shields:

[1] A class of Markov shifts which are Bernoulli shifts, Advances in Mathematics, vol. 6, pp. 323-328, 1971.

L. D. Meshalkin:

[1] A case of isomorphism of Bernoulli schemes, Doklady Akademii Nauk SSSR, vol. 128, pp. 41-44, 1959.

V. Nemytskii and V. Stepanov:

[1] Qualitative Theory of Differential Equations, Princeton, 1960.

J. von Neumann:

[1] Proof of the quasi-ergodic hypothesis, Proceedings of the National Academy of Science USA, vol. 18, pp. 263-266, 1932.

[2] Zur Operatorenmethode in der klassichen Mechanik, Annals of Mathematics, vol. 33, pp. 587-642, 1932.

D. Newton:

[1] On sequence entropy I, II, Mathematical Systems Theory, vol. 4, pp. 119-128, 1970.

193

Z. Nitecki:

[1] Differentiable Dynamics, M.I.T. Press, 1971.

D. S. Ornstein:

[1] Bernoulli shifts with the same entropy are isomorphic,
     Advances in Mathematics, vol. 4, pp. 337-352, 1970.

[2] Two Bernoulli shifts with infinite entropy are isomorphic,
     Advances in Mathematics, vol. 5, pp. 339-348, 1970.

[3] Factors of Bernoulli, shifts are Bernoulli shifts, Advances
     in Mathematics, vol. 5, pp. 349-364, 1970.

[4] Imbedding Bernoulli shifts in flows, contributions to
     ergodic theory and probability, Lecture Notes in Mathema-
     tics, Springer Berlin, pp. 178-218, 1970.

[5] Some new results in the Kolmogorov-Sinai theory of entropy
     and ergodic theory, Bulletin of the American Mathematical
     Society, vol. 77, pp. 878-890, 1971.

[6] An example of a Kolmogorov automorphism that is not a
     Bernoulli shift, Advances in Mathematics, vol. 10, pp. 49-
     62, 1973.

[7] The isomorphism theorem for Bernoulli flows, Advances in
     Mathematics, vol. 10, pp. 124-142, 1973.

[8] A K-automorphism with no square root and Pinsker's conjec-
     ture, Advances in Mathematics, vol. 10, pp. 89-102, 1973.

[9] A mixing transformation for which Pinsker's conjecture
     fails, Advances in Mathematics, vol. 10, pp. 103-123, 1973.

D. S. Ornstein and P. C. Shields:

[1] An uncountable family of K-automorphisms, Advances in
     Mathematics, vol. 10, pp. 63-88, 1973.

[2] Mixing Markov shifts of kernel type are Bernoulli, Advances
     in Mathematics, vol. 10, pp. 143-146, 1973.

D. S. Ornstein and B. Weiss:

[1] Geodesic flows are Bernoullian, Israel Journal of Mathema-
     tics, vol. 14, pp. 184-197, 1973.

J. C. Oxtoby:

[1] Ergodic sets, Bulletin of the American Mathematical Society,
     vol. 58, pp. 116-136, 1952.

W. Parry:

[1] Symbolic dynamics and transformations of the unit interval, Transactions of the American Mathematical Society, vol. 122, pp. 368-378, 1966.

[2] Compact abelian group extensions of discrete dynamical systems, Z. Wahrscheinlichkeitstheorie, vol. 13, pp. 95-113, 1969.

[3] Entropy and Generators in Ergodic Theory, Benjamin, 1969.

[4] Ergodic properties of affine transformations and flows on nilmanifolds, American Journal of Mathematics, vol. 91, pp. 757-771, 1969.

[5] Spectral analysis of G-extensions of dynamical systems, Topology, vol. 9, pp. 217-224, 1970.

[6] Dynamical systems on nilmanifolds, Bulletin of the London Mathematical Society, vol. 2, pp. 37-40, 1970.

M. S. Pinsker:

[1] Dynamical systems with completely positive and zero entropy, Doklady Akad. Nauk SSSR, vol. 133, pp. 1025-1026, 1960 (Russian), Soviet Mathematics Doklady, vol. 1, pp. 937-938, 1960 (English).

L. S. Pontrjagin:

[1] Topological Groups, Gordon and Breach, 1966.

W. Reddy:

[1] The existence of expansive homeomorphisms on manifolds, Duke Mathematical Journal, vol. 32, pp. 627-632, 1965.

V. A. Rohlin:

[1] A general measure-preserving transformation is not mixing, Doklady Akademii Nauk, vol. 60, pp. 349-351, 1948.

[2] Generators in ergodic theory, Vestnik Leningradskogo Universiteta, vol. 18, no. 1, pp. 26-32, 1963.

[3] Selected topics in the metric theory of dynamical systems, American Mathematical Society Translations, Series 2, vol. 49, pp. 171-240, 1966.

[4] Metric properties of endomorphisms of compact commutative groups, Izvestija Akademii Nauk, Serija Matematiceskaja, vol. 28, pp. 867-874, 1964 (Russian); American Mathematical Society Translations, Series 2, vol. 64, pp. 244-252, 1967 (English).

[5] Lectures on the entropy theory of transformations with invariant measure, Uspehi Matematiceskih Nauk, vol. 22, no. 5, pp. 3-56, 1967 (Russian); Russian Mathematical Surveys, vol. 22, no. 5, pp. 1-52, 1967 (English).

[6] Metric properties of endomorphisms of compact commutative groups, Izvestija Akademii Nauk, Serija Matematiceskaja, vol. 13, pp. 329-340, 1949 (Russian).

V. A. Rohlin and Ja. G. Sinai:

[1] Construction and properties of invariant measurable partitions, Doklady Akademii Nauk SSSR, vol. 141, pp. 1038-1041, 1961 (Russian); Soviet Mathematics, vol. 2, pp. 1611-1614, 1961 (English).

D. Ruelle:

[1] Statistical mechanics on a compact set with Z^ν action satisfying expansiveness and specification, Transactions of the American Mathematical Society, vol. 185, pp. 237-252, 1973.

P. C. Shields:

[1] Bernoulli shifts are determined by their factor algebras, Proceedings of the American Mathematical Society, vol. 41, pp. 331-332, 1973.

[2] The Theory of Bernoulli Shifts, University of Chigago Lecture Notes, 1974.

C. P. Simon:

[1] Instability in $\text{Diff}^r(T^3)$ and the nongenericity of rational zeta functions, Transactions of the American Mathematical Society, vol. 174, pp. 217-242, 1972.

Ja. G. Sinai:

[1] On the concept of entropy of a dynamical system, Doklady Akademii Nauk SSSR, vol. 124, pp. 768-771, 1959.

[2] On flows with finite entropy, Doklady Akademii Nauk SSSR, vol. 125, pp. 1200-1202, 1959.

[3] On a weak isomorphism of transformations with an invariant, Doklady Akademii Nauk SSSR, vol. 147, pp. 797-800, 1962 (Russian), Soviet Mathematics Doklady, vol. 3, pp. 1725-1729, 1962 (English).

[4] On a weak isomorphism of transformations with an invariant measure, Matematiceskii Sbornik, vol. 63, pp. 23-42, 1964 (Russian).

196

[5] Classical dynamical systems with countably-multiple Lebesgue spectrum, II, Izvestija Akademii Nauk SSSR, Serija Matematiceskaja, vol. 30, pp. 15-68, 1966 (Russian); American Mathematical Society Translations, Series 2, vol. 68, pp. 34-88, 1968 (English).

[6] Construction of Markov partitioning, Funkcional'nyi Analiz i ego Priloženija, vol. 2, pp. 70-80, 1968.

[7] Markov partitions and U-diffeomorphisms, Functional Analysis and Applications, vol. 2, pp. 61-82, 1968 (English); Funkcional'nyi Analiz i ego Priloženija, vol. 2, pp. 64-89, 1968 (Russian).

S. Smale:

[1] Differentiable dynamical systems, Bulletin of the American Mathematical Society, vol. 73, pp. 747-817, 1967.

M. Smordinsky:

[1] On Ornstein's isomorphism theorem for Bernoulli shifts, Advances in Mathematics, vol. 10, pp. 1-9, 1973.

[2] Ergodic Theory, Entropy, Springer Lecture Notes, no. 214, 1971.

B. Weiss:

[1] The isomorphism problem in ergodic theory, Bulletin of the American Mathematical Society, vol. 78, pp. 668-684, 1972.

Index

198

Lecture Notes in Mathematics

continuation on page 201